T0135652

Dynamics and Ferroelectric Order in Pyridinium Salts, and Host-Guest Interaction of Benzene on Faujasite

Von der Fakultät Chemie der Universität Stuttgart zur

Erlangung der Würde eines Doktors der Naturwissenschaften

(Dr. rer. nat.) genehmigte Abhandlung

vorgelegt von

Dipl.-Chem. Bettina Beck

aus Ibbenbüren

Hauptberichter: Prof. Dr. Emil Roduner

Mitberichter: PD Dr. Günter Majer

Tag der mündlichen Prüfung: 21.03.2003

INSTITUT FÜR PHYSIKALISCHE CHEMIE
DER UNIVERSITÄT STUTTGART

März 2003

Universität Stuttgart
Institut für Physikalische Chemie
Dissertation März 2003

Bibliografische Information Der Deutschen Bibliothek

Die Deutsche Bibliothek verzeichnet diese Publikation in der Deutschen
Nationalbibliografie; detaillierte bibliografische Daten sind im Internet über
http://dnb.ddb.de abrufbar.

ISBN 3-8325-0296-3

Logos Verlag Berlin
Comeniushof, Gubener Str. 47,
10243 Berlin
Tel.: +49 030 42 85 10 90
Fax: +49 030 42 85 10 92
INTERNET: http://www.logos-verlag.de

to Sandra

Contents

Chapter 1

INTRODUCTION

Zeolites are porous, crystalline aluminosilicate framework structures of a regular pore and/or channel system accomodating exchangeable extra framework cations. They are used in detergents and separation processes, but they are also of great importance as catalysts, mostly in the petrochemical industry for the interconversion of hydrocarbons.[1] This is demonstrated in the activities of companies like Exxon Mobile or SHELL in zeolite research and in the number of patents in that field. Especially transition metal exchanged zeolites are interesting catalysts for a variety of chemical reactions.[2–4] Besides the framework structure of the zeolite itself an important question related to the performance of such catalysts concerns the location, oxidation state and distribution of the extra framework transition metal species in the zeolite voids accessible to reactant molecules.[5] It is well known that only species in certain oxidation states and in well defined geometrical and chemical environment within the aluminosilicate framework of zeolites provide catalytic activity. In addition the translational and reorientational mobility of the reactants and their interactions with the cations located in the zeolite framework are of paramount interest with respect to catalytic activity and selectivity. Mostly uncharged molecules or ionic species were investigated, although it is well-known that radicals play an important role as intermediates in catalytic processes.

Zeolite frameworks loaded with adsorbate molecules are quite typical representatives of host-guest systems which were always of paramount interest with respect to translational and reorientational dynamics. Such host-guest systems are not the only examples for solid-state dynamics. It is well known that molecules inside molecular crystals perform rotational motion as well as far as they reveal a symmetry that minimizes steric constraints. This is surprising as a crystal shows a long-range order, *i.e.*, atoms and molecules are situated at defined sites, and they are fixed; nevertheless the crystal is not destroyed although its building blocks perform fast reorientational motion.

Several examples of molecular crystals are found in the family of pyridinium salts. Among these there are compounds built from pyridinium cations and simple an-

ions like chloride, bromide or iodide[6-9] on the one hand and highly symmetric
anions like tetrafluoroborate,[10-14] perchlorate,[15-17] iodate,[18] perrhenate,[19,20] hex-
afluoroantimonate[21] and hexafluorophosphate[22-24] on the other hand. All these
pyridinium salts show at least one solid-solid phase transition; at temperatures
above this transition both the cation and the anion perform a rotational motion,
and even below the transition a certain dynamics remains. The high temperature phase of all the pyridinium salts is paraelectric. The compounds containing
tetrahedral anions and the fluorosulfonate[25] turn out to be ferroelectric in the
low temperature phase. This is thought to be due to a parallel alignment of the
electric dipole moments of the pyridinium cations. Up to now it is not clear why
some of these phase transitions are of first order, whereas others are of second
order although the structural conditions are comparable. In addition there is only
a limited understanding of what the factors are which drive phase transitions. To
describe and understand phase transitions properly detailed information about
the ionic dynamics and the time scale of the dynamics are necessary. Furthermore,
the correlation between ferroelectric polarization and dynamics as well as the effect of steric conditions on this dynamics is of interest to understand the different
behavior of the various pyridinium compounds. Ferroelectric materials are investigated with respect to their application as electronic memory and sensors;[26,27]
but they can also give general information on intermolecular interactions in solids.

One technique to investigate molecular dynamics is ^{2}H NMR. It is based on the
fact that the deuterium atom has a nuclear spin as well as an electric quadrupole
moment. The interaction of the latter with electric field gradients in the bond
dominates the solid-state NMR spectrum of deuterium relaxation; deuterium
therefore represents an excellent probe to investigate the reorientational dynamics
of a molecule.[28] Because the quadrupolar interaction is a single nucleus property,
few structural assumptions are required to analyse the line shape or relaxation.
In addition, deuterium is a nonperturbing probe. The replacement of the protons
with deuterium has a negligible effect on the structure and dynamics. Furthermore, different types of motion can be distinguished through the analysis of the
line shape, and the dynamic range of ^{2}H NMR is large.[29,30]
Another suitable probe to investigate dynamics is the positive muon μ^+ which
does not carry a quadrupole moment like deuterium but a hyperfine tensor in the
radical which is observed. Used in muon spectroscopy it provides an interesting
tool with the muon being a spy on the one hand giving information about dynamics and structure and forming the radical species automatically on the other
hand.[31-34] The μSR technique is a variant of magnetic resonance with respect
to theory and interpretation of the data, but there are distinct differences concerning the production of radicals and the detection method. The advantage of
the muon is that it can be implanted in any material including the ones which
contain no other suitable NMR-active nuclei. For organic muonated radicals the

binding site can be derived with confidence by analogy with the known chemistry of hydrogen atoms. Structure and properties of muonated radicals are close to those of their hydrogen analogs and therefore well understood. But implanting the muon, $e.g.$, into an organic molecule generates a radical which certainly differs in electronic structure from the parent diamagnetic molecule. Muon beams are available with a spin polarization close to 100%. This is an advantage compared to other magnetic resonance techniques. Due to the life time of the muon the frequency resolution and the accessible time window are limited compared to NMR and EPR. The high spin polarization of the muon beam and a single event detection technique used in μSR lead to a high sensitivity.[35]

In the present context a special μSR technique, the ALC-μSR,[36-38] was used to investigate muonated cyclohexadienyl radicals in a series of transition metal exchanged zeolites NaY and NaX. It seems to be reasonable to chose benzene as the radical parent molecule as both, the benzene and the derived cyclohexadienyl radical, are well known from μSR and from different other methods even in zeolite matrices. Benzene is expected to give only one radical species, which simplifies the analysis of the data. In addition, it is assumed that radicals might play an important role as intermediates in catalytic processes. Therefore, their investigation seems to be efficient. And here μSR is quite suitable as the technique itself provides the radical species automatically. The transition metals taken into account in this study are platinum, palladium, silver, nickel and zinc in NaY and manganese in NaX. Although some of them are of minor importance in catalysis they are well suited for basic studies aiming at the determination of the interaction of the radical with the transition metal cation. To enable a direct comparison the pure zeolites NaY and NaX loaded with benzene were investigated as well.[39] It was one aim of this work to show the interaction of the different metal cations with the cyclohexadienyl radical and to achieve information about the radical reorientational dynamics in the zeolite lattice. The metals were chosen with respect to cation charge and to electron configuration to enable a comparison between mono- and divalent ions as well as between para- and diamagnetic species.

The second aim of the present study was to investigate the cation dynamics in pyridinium tetrafluoroborate showing two continuous phase transitions and pyridinium perchlorate revealing two discontinuous transitions. Both compounds were investigated by muon spin resonance; pyridinium tetrafluoroborate in addition by ^{2}H NMR. Special interest lies in the effect of the different order of the phase transitions on the spectra. The μSR technique only sees the pyridinium cation as the muon adds to unsaturated compounds which in this case is the pyridinium ring. In ^{2}H NMR only the deuterons are detected which in the present context are located only in the pyridinium cation as well. Consequently, both methods enable us to analyse only the cation contribution to several physical properties. When the results are compared with the corresponding macroscopic data it is thus possible to determine the individual ionic contribution.

Chapter 2

BASICS

2.1   Muon Spin Resonance (μSR)

In 1937, Neddermeyer and Anderson observed the first unstable elementary particle, the muon, coming from cosmic rays.[40] The muon carries spin $I = \frac{1}{2}$, and its mass amounts to $\frac{1}{9}$ of the proton mass, synonymous to 207 times the electron mass. Its magnetic moment is 3.18 times larger than the magnetic moment of the proton. The muon decays with a mean lifetime of 2.197 μs into a positron, a neutrino and an anti-neutrino under spin conservation with the positron being preferentially emitted in the muon spin direction. It took until 1957 that the bound state of the muon with an electron, the muonium atom was detected: $\mu^+ + e^- \equiv$ Mu.[41] Since its reduced mass and its ionization potential are within 0.5% the same as those of hydrogen, its chemical behavior is also the same as that of H. Only the lower mass leads to different dynamics and thus to a kinetic isotope effect. In a chemical sense Mu can be viewed as a light hydrogen isotope. Like hydrogen, muonium reacts with unsaturated systems and forms radical species, for example cyclohexadienyl radicals from benzene. This was predicted by Brodskii[42] already in 1963, but it took another 15 years to observe this reaction directly in high magnetic fields.[43]

When Garwin *et al.* performed an experiment to prove parity non-conservation in pion and muon decay[44] they observed that the initial amplitudes and the relaxation rates of the muon precession signals depend on the nature of the stopping medium. This was the origin of a new analytic method, the Muon Spin Resonance (μSR) technique.[45] As the muon can be implanted in any environment, it is used to probe the local magnetic field, either as a static spectator or by sampling a certain region as a mobile species.[35] Several experimental techniques using muons have been developed. They are all dubbed μSR where μS stands for *muon spin* and R for *resonance, rotation* or *relaxation* depending on the details of the corresponding technique. The basic principle of all experiments involving positive muons is to irradiate the sample with an energetic spin polarized muon beam from a suitable accelerator. All techniques (except the Avoided Level Crossing technique which is time integrated, for details see below) permit the detection of time evolution of the muon spin polarization as a function of external and/or internal magnetic field. Thus it is possible to differentiate between muons in diamagnetic environment, in Mu and in muonated free radicals and to follow their interconversion.[32]

Due to its low mass and low linear energy transfer during its thermalization process the muon causes relatively little damage near its stopping site. Nevertheless, it can probe radiation chemical effects near the end point of its thermalization track.[35] The negative muon μ^- is also used in science, but only a very limited number of μ^- applications have appeared in literature.[46–48]

The Cyclohexadienyl Radical

In the present context the addition of muonium to benzene, forming a cyclohexadienyl radical, is of relevance. In this radical the muon is substituted in the methylene group as a polarized spin label:

$$C_6H_6 + Mu \longrightarrow {}^\bullet C_6H_6Mu$$

The cyclohexadienyl radical differs from benzene only by an additional hydrogen atom, which in the muonated form is the light hydrogen isotope muonium. The radical is an open shell species, but apart from this it is planar like benzene and only slightly elongated in shape; thus, unless there is a specific chemical interaction due to the unpaired electron, one would expect a very similar dynamic behavior as for benzene. In addition a large number of singly and multiply substituted muonated cyclohexadienyl radicals were observed so that substituent effects on the muon coupling constant are well understood.[32] They follow the same trends as those of H-substituted analogues, and there is a strong correlation between the muon coupling constant of the cyclohexadienyl radical and the proton coupling constant of the α-proton in substituted benzyl radicals, which demonstrates the isoelectronic character of the two species.[49]

The cyclohexadienyl radical has been used to study the structures of substituted cyclohexadienyl radicals,[50] their reaction kinetics,[32] their translational diffusion on the surface of spherical silica grains,[51] their reorientational dynamics in high-silica ZSM-5,[49] and their interaction with copper ions in ZSM-5.[52] Here, it permits the investigation of the radical reorientation dynamics and of the interaction of the radical with metal ions in different metal exchanged zeolites NaY and NaX.

Besides, a corresponding cyclohexadienyl-type radical resulting from the addition of muonium to the pyridinium cation in pyridinium salts, the aza-cyclohexadienyl radical, is of interest:

$$C_5NH_6{}^+ + Mu \longrightarrow {}^\bullet C_5NH_6Mu^+$$

To obtain this radical one of the carbon ring atoms in the benzene is replaced by a nitrogen atom, which introduces the positive charge. Consequently, the muon is not only able to add to the methylene group, but also to the nitrogen atom. In principle the dynamic behavior of the aza-radical derived from the pyridinium salt should be comparable to that of the cyclohexadienyl radical derived from benzene.

2.1.1   Avoided Level Crossing Experiment (ALC)

In ALC-μSR the magnetic field and the incoming muon beam are parallel. The technique takes advantage of the avoided level crossings of magnetic energy levels in an external magnetic field where the eigenstates are mixtures of Zeeman states. This leads to oscillation or relaxation of the muon spin polarization and hence to a decrease of its time integrated value (for details see below). The position, the amplitude and the width of those signals give detailed information about the sample compound itself. The longitudinal field technique is not only applied in time integrated mode but can also be used in time differential mode.[31,35]

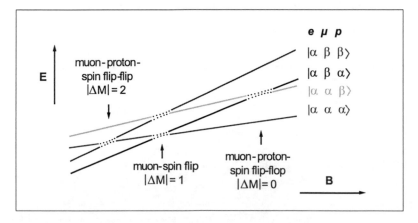

Figure 2.1: Schematic plot of energy levels of a 3-spin-$\frac{1}{2}$-system with electron (e), muon (μ) and proton (p) in high fields. Dotted lines indicate avoided level crossings. Only the states with electron spin α are shown.

Avoided Level Crossings

Regarding a muonated free radical usually three different spin $\frac{1}{2}$ particles have to be considered: the unpaired electron (e), the muon (μ) and a proton (p). In

high magnetic fields this system reveals four different energy levels if only the
muon and the proton are expected to change their state (from α to β: spin flip;
or vice versa: spin flop) (see fig. 2.1). As the four energy levels ($\alpha\alpha\alpha$, $\alpha\alpha\beta$,
$\alpha\beta\alpha$, $\alpha\beta\beta$) do not intersect they show four avoided level crossings. The muon
and the proton spin are able to evolve at such level crossings when reaching the
corresponding magnetic field. A single flip of a muon spin then gives rise to a
transition with selection rule $|\Delta M|=1$, dubbed Δ_1 resonance. It is extremely
sensitive to small anisotropies of the environment and absent in isotropic phases.
This makes it suitable to study radicals in weakly orienting environments, e.g., on
surfaces. In case that the muon and the proton spin change simultaneously but
to different states (muon-proton spin flip-flop) the selection rule $|\Delta M|=0$ leads to
a Δ_0 resonance which is observable in all phases. If muon and proton change to
the same state simultaneously (muon-proton spin flip-flip) the resonance is called
Δ_2 with a selection rule $|\Delta M|=2$. In the presence of hyperfine anisotropy, Δ_1
resonances are normally stronger than Δ_0 resonances. Furthermore, the intensity
of a Δ_0 resonance is related to the product of the muon and the nuclear hyperfine
coupling constant as well as to the resonant field.[32] But the intensity of the
lines is not heavily dependent on the number of nuclei in resonance. In contrast
to conventional magnetic resonance type experiments ALC yields the relative
sign of the coupling constant without special effort. ALC-μSR has proven to be
very sensitive to reorientational dynamics on a critical time scale of the inverse
hyperfine anisotropy, which is typically of the order of 50 ns.

ALC-μSR was applied in this work to determine the reorientatinal dynamics
of benzene adsorbed in various metal exchanged zeolites and of the pyridinium
cations in two different pyridinium salts.

2.1.2   Transverse Field Experiment (TF)

In the transverse field technique the spin of the incoming muon beam is perpendicular to the magnetic field leading to a precession of the muon spin. In a
histogram this precession is displayed as an oscillation. The different oscillation
frequencies relate to the transitions between different magnetic energy levels in
the muon adduct radical. A Fourier transformation of the transverse field μSR
spectra gives directly the precession frequencies of the different muon spins. The
transverse technique is sensitive only to $|\Delta M|=1$ resonances. It is only used in
time differential mode.

In this work transverse field μSR was used solely to identify the Δ_1 resonances
in the ALC-μSR spectra of the pyridinium compounds.

2.2   Nuclear Magnetic Resonance (NMR)

Dynamic NMR techniques have been extensively used to investigate dynamic processes in molecular solids.[30,53–57] Among these, ^{2}H NMR methods have demonstrated a particular suitability in the evaluation of the motional and the structural characteristics on a molecular level.[57–60] Quite different chemical systems such as inclusion compounds, liquid crystals, biological membranes, or polymers have been studied by this means in recent years. Covering several orders of magnitude, various experimental techniques such as line shape analysis and spin-lattice relaxation measurements have been applied to study molecular processes depending on the dynamic range. The corresponding ^{2}H NMR spectra are sensitive to molecular processes with rate constants on the order of the quadrupolar coupling constant, *i.e.*, between 10^4 and 10^8 s^{-1}. Very fast motions with rate constants between 10^8 and 10^{11} s^{-1} can be studied by spin-lattice relaxation measurements (see fig. 2.2).[61] The particular situation in ^{2}H NMR spectroscopy arises from the fact that the spin Hamiltonian is dominated by the quadrupolar interaction with the main interaction axis along the C-^{2}H bond. Consequently, detailed information about actual molecular characteristics of the system under investigation can be provided by the analysis of such ^{2}H NMR experiments.

NMR measurements were used in this context to investigate the reorientational dynamics of the pyridinium cations in pyridinium tetrafluoroborate as a complementary method to the μSR technique.

Table 2.1: Comparison of μSR and NMR.[32]

	TF-μSR	NMR
time resolution	1 ns	1 μs
frequency resolution	0.5 MHz	0.1 Hz
polarization	0.7	10^{-5}
minimum number of spins for a simple spectrum	10^7	10^{17}
detection method	single particle counting	induction coils

2.3 μSR and NMR - a Comparison

A comparison of μSR and NMR by some typical characteristic values given in table 2.1 shows essential differences. The duration of the pulse needed in NMR to tip the magnetization angle limits the time resolution. As in TF-μSR the muon is injected with the spin transverse to the field this pulse is not necessary. But the muon lifetime limits the duration of the FID, which is related to the frequency resolution. Polarization in NMR is usually small because it depends on the Boltzmann population of the energy levels and on temperature. In μSR it is given by the polarization of the muon beam leading to far less spins needed in μSR than in NMR to produce a detectable signal. This effect is even enhanced by the high sensitivity of the single particle counting technique. The different time windows covered by different techniques performed in μSR and NMR are displayed in figure 2.2.

In μSR the widths of the Δ_1 resonances are much more sensitive to reorientational dynamics than the corresponding Δ_0 lines. Rotation faster than the time scale set by $\tau_{ALC} \approx 50$ ns for C_6H_6Mu partially or completely averages out the hyperfine anisotropy, giving rise to motionally narrowed lines. This feature has aspects in common with ^{2}H NMR, but on a much longer time scale of $\tau_{NMR} \approx 10$ μs. Partial averaging of the hyperfine anisotropy, due to fast motion about a preferred axis (uniaxial rotation of C_6H_6Mu), gives rise to the asymmetric shape of the axial powder pattern, in contrast to the symmetric shapes seen for the static limit. In ^{2}H NMR the correspoding situation is the observation of a Pake doublet with reduced quadrupolar splitting. For fast but random isotropic motion, the Δ_1 transition disappears in the μSR case, due to complete motional averaging, whereas in the parallel situation in ^{2}H NMR, the Pake doublet collapses to a single line.[39]

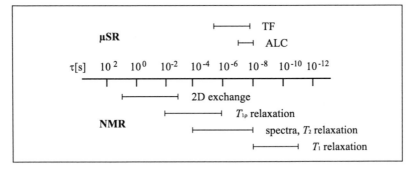

Figure 2.2: Time scale of different techniques in NMR and μSR. τ is the correlation time.

2.4   Zeolites

2.4.1   Characteristics

"A zeolite mineral is a *crystalline* substance with a structure characterized by a *framework of linked tetrahedra*, each consisting of four oxygen atoms surrounding a cation (mostly Si^{4+} and Al^{3+}). This framework contains open cavities in the form of *channels and cages*. These are usually occupied by *water molecules* and *extra framework cations* that are commonly exchangeable. In the hydrated phases, dehydration occurs at temperatures mostly below about 400°C and is largely *reversible*. The framework may be interrupted by OH- or F-groups; these occupy a tetrahedron apex that is not shared with adjacent tetrahedra."[62]

This definition of zeolites was given by Coombs in 1997 and puts the emphasis on the importance of the structure and not on the composition of a material.[62] Zeolites occur in nature and have been known for almost 250 years as aluminosilicate minerals. The swedish mineralogist Crønstedt introduced the name "zeolite" for certain silicate minerals in allusion to their behavior on heating (greek: zeo $\hat{=}$ to boil, lithos $\hat{=}$ stone).[63] Since 1935 systematic investigations on zeolite synthesis are performed. And between 1949 and 1954 the synthesis of zeolites A, X and Y were developed, the latter being up to now the most used zeolitic materials in industry. The fluid catalytic cracking (FCC) of heavy petroleum destillates on zeolite X invented by Mobil Oil in 1962 certainly was a milestone. And between 1967 and 1969 the most important zeolites with low Al-content, ZSM-5 and Beta, were synthesized also by Mobil Oil.

Natural zeolites contain undesired impurity phases, and their chemical composition varies; thus they are only used for low-demanding applications like in agriculture, or as building materials. In clear contrast, synthetic zeolites are high performance materials as it is possible to optimize the structure with respect to the demands. According to the IUPAC zeolites belong to the microporous materials as their pore diameters are smaller than 2.0 nm which is in the order of molecular dimensions. The strictly regular pore architecture guarantees reproducable and well defined characteristics which is, *e.g.*, one requirement for shape selective catalysis. As the extra framework ions are exchangeable zeolites are used as cation exchangers and show an interesting host-guest chemistry. They are stable up to high temperatures, and there are many different methods available for their modification.[4,5,64]

In 1999 the global market for commercial zeolites was more than 1.5 Mio. to.[1] The major application of zeolites with respect to weight is in detergents, followed by the use of natural zeolites and by consumption in catalytic and adsorbent applications. The big advantage of zeolites in catalytic processes is the fact that their properties can often be tailored to suit a particular application. In addition important properties in catalysis are the concentration effect, the shape selectivity, the acidity as well as thermal and hydrothermal stability of zeolitic materials. For the performance of catalytic processes involving zeolites the host-guest interactions and the mobility of the reactants, the intermediates and the products are of paramount importance. This holds for the translational mobility, as reactants have to reach the active sites and products have to leave the porous system. It is the basis for the efficiency of a catalyst. Reorientational mobility is a measure of similar spatial constraints and of specific interactions. For transient intermediates, in particular, it influences transition state selectivity and reaction pathways in general.[2,49]

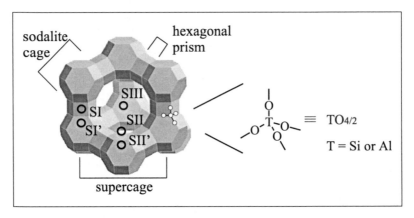

Figure 2.3: Faujasite structure showing the most important cation sites (SI, SI', SII, SII', SIII), the basic building blocks and the tetrahedron. (Drawing of framework structure taken from: www.iza-structure.org/databases/: International Zeolite Association; see also for structural details of any zeolite material.)

Structure

The typical chemical composition of an aluminosilicate zeolite can be described as follows:

$$A_{y/m}^{m+} \left[(SiO_2)_x \, (AlO_2^-)_y \right] \cdot z \; H_2O$$

where the square bracket contains the anionic aluminosilicate framework and A are the extra framework cations charged +m and compensating the negative charge of the AlO_2 tetrahedrons in the framework. The basic element of aluminosilicate zeolites is the $TO_{4/2}$ tetrahedron (see above: in parenthesis) with

T=Si or Al (see figure 2.3). Connecting such tetrahedrons via common oxygen atoms leads to different characteristic building blocks. Depending on the fashion of linkage these building blocks form various threedimensional frameworks with interconnecting channels and/or cages of different size.

One of these building units is the so-called sodalite or β-cage ($\oslash \approx 0.66$ nm) composed from 24 Si or Al and 36 O atoms. Joining the sodalite cages over their six-membered rings gives hexagonal prisms (double six rings = D6R) and additional larger cages, dubbed supercages (in the following: sc) or α-cages ($\oslash \approx 1.3$ nm) leading finally to the faujasite structure (FAU), a threedimensional channel system (see figure 2.3). One unit cell of the faujasite structure comprises eight soldalite cages and eight supercages. Each supercage is joined to four others via circular tetrahedrally arranged 12-ring windows ($\oslash \approx 0.74$ nm) formed by 12 Si or Al atoms and 12 bridging O atoms. Compared to other zeolite structures this opening is quite large. Thus the faujasite type zeolite belongs to the group of large-pore zeolites. Zeolitic materials with pore openings larger than a 12-ring window are very seldom.[5,65]

The defined pore and/or channel structure of the zeolites is the basis for one of the most important characteristics of these materials which is called the molecular sieve effect. Depending on the different size and different shape of pore openings only specific molecules can enter the pore structure and are ad- and desorbed. This enables a wide field of applications for zeolites as selective adsorbates in separation processes and as shape selective catalysts. As the extra framework cations are exchangeable and usually very mobile zeolites are used as well as ion exchangers.

According to Loewenstein's rule Al–O–Al linkages are forbidden.[66] Consequently the ratio of the Si content and the Al content is always bigger than or equal to unity ($x/y = n_{Si}/n_{Al} \geq 1$). Materials with x/y =1-5 are called aluminium rich, with $x/y = 20$-∞ silicon rich zeolites. The aluminium content of a zeolite material has a direct influence on many of its characteristics, $i.e.$, the charge density of the framework, the ion exchange capacity, the density and the strength of the acidic sites and the polarity of the surface.

The Faujasites NaX and NaY

Synthetic faujasite materials are mainly zeolite X and zeolite Y. The unit cells of X and Y are very large (nearly 25 Å) and the framework contains the largest void space of any known zeolite; it amounts to about 50% of the dehydrated crystal. The framework structre itself is remarkably stable and rigid. Although X and Y have topologically the same aluminosilicate framework structure they show characteristic differences. They are distinguished on the basis of chemical composition, which is related to the synthesis method, structure, and their related physical and chemical properties. Differences are found in the cation composition and distribution, the Si/Al ratio and possible Si-Al ordering in the tetrahedral sites. The number of aluminium ions in the unit cell of zeolite X varies from 96 to about 77. In zeolite Y it is about 76 to 48. The Si/Al ratio is x/y = 1-1.5 for zeolite X and x/y = 1.5-3 for zeolite Y. Both, X and Y, are usually synthesized in the sodium form NaX and NaY.[5] The unit cell composition of the two zeolites used in this work is as follows:

NaX: $Na_{80}[(Si_{112}O_{224})(Al_{80}O_{160})] \cdot 240\ H_2O$ x/y = 1.4

NaY: $Na_{58}[(Si_{134}O_{268})(Al_{58}O_{116})] \cdot 250\ H_2O$ x/y = 2.3

One of the most important properties is the distribution and therefore the accessibility of the extra framework cations. The cation sites and their designation in zeolites X and Y are described in the following, starting in the smallest opening, the hexagonal prism and proceeding over the sodalite cage to the largest opening, the supercage (see also figure 2.3): site I is located in the center of the double 6-ring (D6R = hexagonal prism), whereas site SI' is on the inside of the β-cage adjacent to the D6R. Site SII' is as well inside the sodalite cage but adjacent to the single 6-ring, i.e., next to the supercage. Site SII approaches the single 6-ring outside the sodalite unit, i.e., in the large cavity, opposite to site SII'. Site SIII refers to positions in the wall of the supercage, on the 4-fold axis in the 12-ring aperture. The designated sites are only fractionally occupied by cations, depending on the Si/Al ratio and the hydration state. In addition, e.g., sites SI and SI' or SII and SII' cannot be occupied simultaneously as their distance is too short. Consequently the ion exchange capacity of zeolite materials diverges. In addition not all of the sites are available for every type of ion to be exchanged as the ion has to pass different apertures of the different cages in the zeolite structure. A hydrated zeolite NaX with x/y = 1.4 contains 16 Na^+ in SI and 32 in SII, while the remaining 32 ions have not been located and are believed to be hydrated and mobile. No crystallographic study of hydrated NaY has been reported; but in dehydrated NaY 30 cations are located at site SII, 20 at site SI' and 8 at site SI.[2,5]

In order to replace cations located in the small cages of zeolite structures the ions have to pass through a 6-membered ring. Most hydrated metal cations can easily be stripped off at least a part of their hydration shell temporarily, so that they can be ion-exchanged even into small-pore zeolites. On the other hand, elements that do not readily form hydrated cations, e.g., Pt, are usually ion exchanged in the form of their ammine complexes, i.e., $[(NH_3)_4Pt]^{2+}$. The largest univalent ions that can completely replace Na^+ in a faujasite structure are expected to be K^+, Ag^+ and Tl^+.

To introduce other elements into the pore system of zeolites, ion exchange procedures are used; examples are the exchange of Ca and K into zeolite A and Y, rare earths (La, Sm, Ce, etc.) into zeolites X and Y, and Pt into mordenite and zeolite Y. Many cationic elements can be introduced by the conventional *excess solution* procedures. Here, the cationic salts are dissolved in enough solvent to simply fill the pores of the zeolite. Nitrates, sulfates, chlorides, or organic salts such as acetates are often used in zeolite exchanges because of their solubility in water and because they are easily decomposed by heating.

2.4.2   Benzene in Zeolites

Reorientational dynamics of diamagnetic molecules in zeolites has been studied mainly by means of nuclear magnetic resonance[7,67–72] and by quasi-elastic neutron scattering,[73] but Monte Carlo and molecular dynamics calculations[69,74] have contributed much to a detailed understanding. To reduce complexity, investigations have concentrated on simple molecules of high symmetry and on a few well-defined zeolite structures. One of the best known systems is that of benzene in silicalite I.[49] In addition the interpretation of catalytic behavior requires a knowledge not only of the adsorbate reorientational dynamics but also of the cation distribution within the zeolite cavities and, mainly, of the interactions between the cations and the adsorbed molecules, the framework and the adsorbed molecules and as well between the molecules themselves.

The interaction of benzene with sodium cations in zeolite NaY and NaX has been investigated by various techniques.[69,75] Two important adsorption sites for benzene in NaY have been reported in a low temperature powder neutron diffraction study for a temperature of 4 K. At room temperature the benzene molecules are clearly delocalized but still located close to the sites determined at 4 K.[65] The first site, which is thought to be occupied preferentially, is found in the supercage, where it is facially coordinated to the SII sodium ions located at the six-ring windows connecting the super- and the β-cages. The second benzene site is located in the 12-ring windows that link neighboring supercages and provide a well tailored environment for the benzene molecule; this second adsorption site becomes more important at higher concentrations of benzene, *i.e.*, 2.5 molecules/sc.[68] These results have been verified only recently by Fleming and coworkers for muonated cyclohexadienyl radicals, generated via the muonium addition to benzene, using Avoided Level Crossing Muon Spin Resonance. In addition to the above mentioned adsorption sites they found that the radical is able to coordinate to the sodium cations at site SII in two different orientations, with the muon in exo or in endo orientation with respect to the sodium ion. This results in large shifts of the hyperfine coupling constants showing the strong bond formed between the radical and the sodium cation.[39] There is a maximum of six sites available per supercage (4/sc at the first site, 2/sc at the second site); but neutron scattering studies have suggested that at high benzene coverages ($\approx$2.5/sc) there is a ten-

dency for the benzene molecules to aggregate in a limited number of cavities to
form clusters, whereas at low coverages ($\approx 1.0/sc$) most of the benzene molecules
are rather evenly distributed throughout the channels in the zeolite.[76] The effect of benzene adsorption on the faujasite framework is small. Nevertheless,
at high coverages, a sodium ion displacement due to the benzene molecules has
been detected.[65] For NaLSX and CaLSX the same adsorption sites for benzene
were identified.[68] The heat of adsorption of benzene in NaY and NaX is known
to increase as the Si/Al ratio decreases, reflecting the dominating influence of
cation sorbate interactions. The benzene molecule, when adsorbed in a zeolitic
structure, is well known to perform fast reorientation about the sixfold axis in a
temperature range between room temperature and 100 K.[69] In addition at higher
temperatures jumps between different adsorption sites were detected depending
on the framework, the temperature and the loading.[?,71,72] In NaY the diffusion
coefficient of benzene is much smaller and its activiation energy is higher than
in, $e.g.$, silicious faujasite, in which no cations are present, indicating once more
the strong interactions between the adsorbate and the cation.[69] Reorientational
motion of cyclohexadienyl radicals accomodated in zeolite NaY and interacting
with the sodium cations was excluded by Macrae. Quantum chemical calculations showed that the radical is rigidly fixed due to its dipole moment.[77]

A very interesting feature is the interaction of adsorbate molecules with transition metal cations like platinum or palladium in zeolitic matrices as those metal-
exchanged zeolites are used in catalytic processes in the petroleum refining industry.[78] Roduner $et\ al.$ investigated such a system, $i.e.$, a complex of a neutral organic free radical with a diamagnetic metal ion in a zeolite. They determined the
interaction of muonated cyclohexadienyl radicals with copper ions in CuZSM-5
which is an important catalyst for NO decomposition. The radical was derived
from benzene via muonium addition in Avoided Level Crossing Muon Spin Resonance. Roduner $et\ al.$ observed clearly an interaction of the radical with the
copper ions and considerably less motional averaging of the adsorbed molecules.
The complex is believed to be located at the channel intersections of the zeolite
due to spatial constraints.[49]

2.5   Pyridinium Salts

An interesting family of pyridinium compounds is easily formed by a reaction of
pyridine, a strong organic base, with many acids. The majority of these salts
undergo a single structural solid-solid phase transition of the order-disorder type
changing the cation dynamics.[6–8,21,23] In the high temperature phase the pyridinium cation performs fast rotation around the pseudo $C6$ axis perpendicular to
the molecular plane between equivalent potential wells, whereas in the low tem-

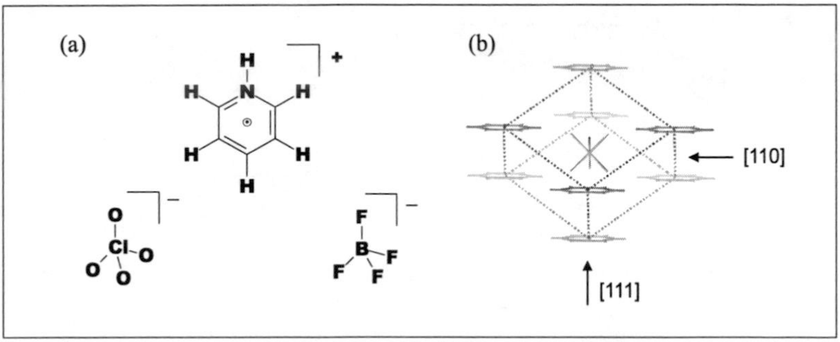

Figure 2.4: Cation and anion structure of $PyBF_4$ and $PyClO_4$ (a). Schematic unit cell of $PyBF_4$ of the high temperature phase (b). The structural changes at the phase transitions are minute.

perature phase the rotation takes place between inequivalent potentials. Only some of the compounds show a sequence of two transitions. The additional transition always takes place at higher temperatures and is ferroelectric.[15, 18–20, 25]

2.5.1 $PyBF_4$

With reference to Czarnecki *et al.* the very outstanding example of this family of pyridinium salts is pyridinium tetrafluoroborate ($PyBF_4$), as it is the only multiaxial ferroelectric undergoing a continuous phase transition. On cooling, the crystal transforms from $R\bar{3}m$ to $C2$ symmetry at 238.7 K, and at 204 K to the space group $P2$. The transition at higher temperature is paraelectric-ferroelectric and of second order according to Czarnecki *et al.*[10, 11] In clear contrast Hanaya *et al.* are convinced that this transition must be of first order due to a different interpretation of the characteristics of the heat capacity anomalies.[79] The cation and anion structures as well as the schematic unit cell of the high temperature phase of $PyBF_4$ are shown in figure 2.4.

2.5.2 $PyClO_4$

Another member of the family of pyridinium salts is pyridinium perchlorate ($PyClO_4$). It is also a multiaxial ferroelectric undergoing two solid-solid phase transitions of the order-disorder type. But in contrast to the tetrafluoroborate these transitions are of first order.[15] On cooling, the crystal transforms from $R3m$ to Cm symmetry at 248 K, and at 233 K to the space group Cm or Pm. The transition at higher temperature is once again paraelectric-ferroelectric. The cation and anion structure of $PyClO_4$ are shown in figure 2.4.

2.6   Phase Transitions

2.6.1   Theories

Ehrenfest Classification

Thermodynamic systems can exist in various stable homogeneous states, called phases, which might differ in structure, symmetry, order and dynamics. At certain values of the variables of state (like temperature or pressure) the state of the system changes, a phase transition takes place.[80] Many different types of phase transitions are known, including the more common ones like fusion and vaporization and less common examples like solid-solid, conducting-superconducting, ferromagnetic-paramagnetic, or fluid-superfluid transitions. It is possible to classify phase transitions into different types using the behavior of the chemical potential known as the Ehrenfest classification of phase transitions. If the first derivatives of Gibbs free energy with respect to pressure and temperature given by

$$S = -\left(\frac{\delta G}{\delta T}\right)_p \qquad \text{and} \qquad V = \left(\frac{\delta G}{\delta p}\right)_T \qquad (2.1)$$

are discontinuous, the transition is classified as first order. This means that the entropy and the volume show an abrupt, finite, nonzero change ($\Delta_{tr}S \neq 0$, $\Delta_{tr}V \neq 0$) at the specific transition temperature. The discontinuous behavior of the entropy and the volume thus leads to infinite values of the heat capacity

$$c_p = T\left(\frac{\delta S}{\delta T}\right)_p = -T\left(\frac{\delta^2 G}{\delta T^2}\right)_p \qquad (2.2)$$

and the thermal expansion coefficient

$$\alpha = \frac{1}{V}\left(\frac{\delta V}{\delta T}\right)_p = \frac{1}{V}\left(\frac{\delta}{\delta T}\left(\frac{\delta G}{\delta p}\right)_T\right)_p. \qquad (2.3)$$

Common examples of first order phase transitions are fusion and vaporization. Characteristic phenomena which might accompany a first order phase transition are supercooling and superheating, *i.e.*, passing the phase transition temperature without changing the state of the system immediately. In contrast, for second order transitions the first derivatives of the Gibbs energy function are continuous ($\Delta_{tr}S = 0$, $\Delta_{tr}V = 0$), but the second derivatives of G are discontinuous now. That is why the heat capacity and the thermal expansion now show finite jumps at the phase transition temperature. A normal-to-superconducting phase transition at zero applied magnetic field is an example of a second order transition. A second order phase transition does not exhibit supercooling nor superheating.

Third and higher order phase transitions can be described in an analogous fashion. A phase transition that is not first order but where the heat capacity becomes nevertheless infinite is called a λ-transition. Here the shape of the heat capacity curve resembles the Greek letter *lambda* as the curve already starts increasing well before the transition temperature. Examples for λ-transitions are order-disorder transitions in alloys, the onset of ferromagnetism, and the fluid-superfluid transition of liquid helium.[81,82]

Landau's Theory

Ehrenfest himself realized that his classification was not sufficient to explain the basic differences between first and higher order phase transitions. Landau was the first scientist who noticed that during a second order phase transition the symmetry of the system *must* change. Thereby, the ordered phase belongs to a symmetric subgroup of the disordered phase. Although parameters like volume and entropy might change continuously with pressure or temperature, the symmetry can only change discontinuously. This was the basis of Landau's theory of second order phase transitions. Landau introduced an order parameter η, with $\eta = 0$ for a disordered and $\eta \neq 0$ for an ordered state. The symmetry of a body is only changed when η becomes identically zero. In general, but not without exception, the ordered state is the low temperature phase, whereas the disordered phase is to be found in the high temperature regime. Some dedicated conclusions concerning the essential differences of first and second order phase transitions are known as Landau rules and summarized in table 2.2. To consider the thermodynamic consequences, Landau proposed the expansion of the thermodynamic potential, *e.g.*, the Gibbs free energy G, in the vicinity of the phase transition temperature in powers of the order parameter η:

$$G(p,T,\eta) = G_0 + A\,\eta + \frac{1}{2}\,a\,\eta^2 + \frac{1}{3}\,B\,\eta^3 + \frac{1}{4}\,b\,\eta^4 + \frac{1}{5}\,C\,\eta^5 + \frac{1}{6}\,c\,\eta^6 + \dots \quad (2.4)$$

Taking into account the conditions of thermodynamic equilibrium and stability, regarding that G is invariant under sign inversion and proposing an assumption for the temperature dependence of the expansion coefficients (for details see 80) leads to:

$$G(p,T,\eta) = G_0 + \frac{1}{2}\,\alpha(p)\,(T - T_C)\eta^2 + \frac{1}{4}\,b(p)\,\eta^4 + \frac{1}{6}\,c(p)\,\eta^6 + \dots \quad (2.5)$$

A discussion of equation 2.5 shows that the order of the phase transition is determined by the sign of the fourth order coefficient $b(p)$:

Table 2.2: Summary of Landau rules on first and second order phase transitions.[80]

First order transition	Second order transition
Symmetry may change or may not (no symmetry relation).	Symmetry *must* change. The transition state has to include all symmetry elements of *both* phases (symmetry relation).
Transition point: Both phases coexist with phase boundaries between; superheating and supercooling possible.	Transition point: Both "phases" represent the *same state of matter*. No phase boundaries; superheating and supercooling impossible.
Order parameter varies *discontinuously* with temperature.	Order parameter varies *continuously* with temperature.

$$b(p) > 0: \text{ second order}$$
$$b(p) = 0: \text{ tricritical}$$
$$b(p) < 0: \text{ first order}$$

In case of a second order phase transition including a finite jump of the heat capacity $b(p)$ is positive and the order parameter η varies according to the power law

$$\eta \propto (T_C - T)^{1/2}. \tag{2.6}$$

If $b(p)$ is negative, a first order transition takes place accompanied by a latent heat. The borderline case with $b(p)=0$ is the so-called tricritical transition. The molar heat capacity rises to infinity and the order parameter η varies according to

$$\eta \propto (T_C - T)^{1/4}. \tag{2.7}$$

The Landau expansion 2.5 may be truncated after the fourth order term in the case of second order transitions and after the sixth order term in the case of first order and tricritical transitions.[80]

2.6.2   Ferroelectricity

Certain dielectrics generate an electrical polarization when subjected to an external mechanical stress (or vice versa). Such materials are called *piezoelectric*. A
subset of piezoelectric solids are those in which a spontaneous electrical polarization in a crystal is caused by an intrinsic internal strain accompanying a change
of crystal structure to one of lower symmetry. Such materials are termed *pyroelectric* because the natural spontaneous electrical polarization is usually masked
by neutralizing counter-ions adsorbed onto the free surface; it can thus only be
revealed by heating the sample, which removes some of the counter-ions. A further subset of pyroelectric materials comprises those in which the spontaneous
electrical polarization in a unit cell can be reversibly changed between $\pm P_S$ by the
application of an electric field of suitable polarity. In this way, the polarization
can be aligned in neighbouring domains. These materials are called *ferroelectric*.
The temperature at which the transition takes place between the randomized
paraelectric and the ordered ferroelectric crystal phases is termed the ferroelectric Curie temperature T_{Cf}. In common with other types of phase transitions,
ferroelectric transitions can also be divided into first order and second order (see
chapter 2.6.1). A first order ferroelectric transition is not only characterized by
a discontinuity in volume and entropy but also in P_S at T_{Cf} ; an example is the
cubic-tetragonal transition in $BaTiO_3$. Second order transitions, in contrast, are
those where there is no change in P_S, but $\delta P_S / \delta T$ is discontinuous (as is the
specific heat). $LiTaO_3$ is an example of a material that exhibits a second order
ferroelectric phase transition.

Although for normal dielectrics the electric polarization is proportional to the
macroscopic electric field inside the solid, this behavior does not hold for ferroelectric materials even though reversal of an electric field can reverse the saturation polarization. Instead, hysteresis is observed; the curve for the polarization
for increasing electric field is different from that for decreasing electrical field, and
a hysteresis loop is obtained between the two saturation values of the polarization
$\pm P_S$. The electric field needed to reduce the polarization to zero is called the
coercive field. As a corollary, the polarization remaining at zero field is termed
the remanent polarization. The area inside the hysteresis loop gives the loss in
energy density per cycle.[83]

It is well known that ferroelectricity can be obtained via a displacement of the
different ionic sublattices against each other - the classical ferroelectric mechanism. If one of the ionic sublattices is taken to be fixed the displacement of the
second sublattice leads to a separation of the center of charges and thus to a
polarization which is a requirement for ferroelectricity. The displacement of the
ionic sublattices, which is small compared to the interatomic distances, is always
accompanied by a change in crystal structure. But ferroelectricity is also caused
by an increase in ordering of electrical dipole moments. Random distribution of

dipole moments does not result in a net polarization; but as soon as the dipoles are aligned a net polarization is detectable. Random distribution of dipole moments is achieved, *e.g.*, in carrier molecules performing a fast rotational motion. As soon as this rotation is hindered the dipole moments are no longer averaged out resulting in a net polarization.

Chapter 3

EXPERIMENTAL

EXPERIMENTAL TECHNIQUES, DATA ANALYSIS and SAMPLE PREPA-
RATION

3.1   Muon Spin Resonance

3.1.1   Muons: Production and Decay

Muons occur in nature as a result of several decays in cosmic rays. When light
nuclei, mainly protons from primary cosmic rays, fall on the earth with high
energy they collide with molecules in the upper atmosphere leading to interactions
with their nuclei. Among the new particles produced is the positive pion. It
decays with a mean life time of 26 ns into a positive muon and a neutrino. As the
pion is a spin zero particle and the neutrino shows spin $\frac{1}{2}$ and negative helicity
the muon has to show also spin $\frac{1}{2}$ and negative helicity due to conservation of
angular momentum.

To use muons in the laboratory they are produced in suitable accelerators. For
this purpose electrons or protons are accelerated to energies higher than the pion
rest mass of 140 MeV. After collision with a production target the pions are
emitted and those with the desired charge and momentum are selected. The
pions decay in flight over several meters and deliver a spin polarized muon beam
due to the above mentioned decay characteristics.

$$\pi^+ \longrightarrow \mu^+ + \nu_\mu \tag{3.1}$$

The so-called *decay muons* have a variable momentum typically in the range
between 70 and 130 MeV/c, requiring a stop range of several centimeters of
water. In the case that the pions decay at rest at the surface of the production
target they produce muons with a momentum of 28 MeV/c and a stop range
of 140 mg cm^{-2}, corresponding to a water layer of 1.4 mm thickness. Due to
the birth of the muons at the target surface they are called *surface muons*. The
advantage of surface muons is that they allow working with thinner samples and

lower costs of beam line magnetic elements. In addition, as low energy beams are more easily handeled, histograms are cleaner with less effort. The muon itself decays with a mean lifetime of $\tau_\mu = 2.197$ μs into a positron, a neutrino and an anti-neutrino under spin conservation. The positrons are emitted preferentially in the muon spin direction.[35]

$$\mu^+ \longrightarrow e^+ + \nu_e + \nu_\mu \qquad (3.2)$$

All experimental techniques have in common that they use muons from a spin polarized beam which are stopped in the sample. The sample itself is placed in an external magnetic field. With this experimental setup it is possible to monitor the evolution of the muon spin polarization in the sample via detection of the decay positrons with a single particle counting technique developped by Garwin et al.[44] As the decay positron passes through a scintillator it provides a fluorescence light pulse which is converted to an electrical pulse in a phototube.

Regarding the time structure of the muon beam there are two different types of sources delivering a continuous muon beam or providing a pulsed muon supply. With continuous muon beams the time interval between incoming muon and decay positron can be measured with high precision. But as the last muon has to be decayed before the next muon is allowed this beam type limits the flux to 10^5 muons per second, which is far less than that what modern accelerators can provide. The pulsed sources do not have this limitation, as a large number of muons are allowed simultaneously. In the off-period the decay positrons are accumulated. This leads to a background-free histogram and makes a longer experimental time window accessible. But for the pulsed sources a primary pulse of a length typically of 50 ns is needed, limiting the maximum observable precession frequency to a few Megahertz. Thus, the two types of beam structures provide complementary experimental conditions. The continuous beams are suitable for fast processes and high precession frequencies, whereas the pulsed sources are better for slow processes relying on low background between the pulses. In addition, the pulsed method is good for experiments requiring coincidence with other pulses like laser or radio frequency excitation.[35] The Paul Scherrer Institute (PSI), where all μSR experiments presented in this work were performed, provides a continuous muon beam. ALC experiments were carried out at the πE3 beam port using surface muons, TF measurements at the πE1 beam port using decay muons.

3.1.2   Time Integrated Longitudinal Field μSR (ALC)

The experimental setup of the ALC-μSR measurements is shown in figure 3.1. The incoming muons are stopped in the sample. The muon spin points opposite to flight direction and is parallel to the external magnetic field. After the decay of the muon the positron is emitted from the sample and detected in the backward or the forward counters depending on the previous interactions of the muon within the sample. The ratio of the difference of counted events in the forward and in the backward counter and the total sum of detected events gives the asymmetry As. In case of interactions As shows characteristic resonances with respect to the magentic field.

Samples for ALC measurements are usually kept in metal containers fixed on copper plates connected to the cryostat to enable rapid thermal equilibrium. The muons enter the sample cell through a thin metal window. The thickness of the sample itself has to be sufficient and depends on the density of the compound and the energy of the muons (see also chapter 3.1.1) so that the muons are stopped in the sample compound and cannot pass the cell. These sample cells are appropriate for solids as well as for liquid and for gaseous compounds. For details see chapter 3.4. Measurements are performed under vacuum to avoid spin exchange with oxygen molecules.

To avoid influences of changing beam optics with field, a reference run of a sample of the same design containing the pure zeolite material (no exchanged metal ions, no benzene loading) was used to normalize the spectra. Nevertheless, a general problem of this experimental setup remains the background. Although the ALC data are background corrected they do not show a horizontal base line.

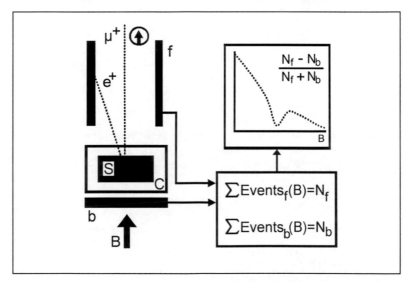

Figure 3.1: Schematic experimental setup of ALC-μSR (f: forward counter, b: backward counter, S: sample, C: cryostat, B: magnetic field, μ^+: incoming muon (arrow indicates the muon beam polarization), e^+: decay positron).

3.1.3   Time Differential Transverse Field μSR (TF)

In transverse field measurements the magnetic field is perpendicular to the muon beam. Backward and forward counter detect the incoming muons and the decay positrons, and a fast clock measures the lifetime of an individual muon in the experimental sample. The experimental setup is in principle equivalent to that shown in figure 3.1. Only the incoming muon spin is rotated by 90°. Due to this setup the muon spin performs precession. For an unpolarized muon beam the number of decay positrons N versus the muon lifetime gives a simple exponential decay with a correlation time τ_c. In contrast, for a polarized muon beam the spin precession results in a modulation of the count rate in the positron detector, owing to the anisotropy in the decay. This is seen as an oscillation on the histogram representing an FID. Interactions of the muons with the sample compound will give changes in this FID. Like in NMR the signals are interpreted as transitions between magnetic states following specific spectroscopic selection rules. The signals in a Fourier transformation are the equivalent of ENDOR transitions and yield directly the muon hyperfine coupling constants.[32] The sample compounds for transverse field measurements are kept in DURAN or Pyrex glass bulbs of approximately 25 mm diameter which are sealed after filling. The bulbs enable to measure solids, liquids and gases.

3.1.4   Data Analysis

Coupling Constants

As described in chapter 2.1.1 three types of resonances are expected to occur for a muon/electron/nuclear system. In the following the calculation of coupling constants from the resonance positions is explained.[49]

The dipolar part of the hyperfine interaction Hamiltonian gives rise to a muon spin flip transition Δ_1 at a resonant field $B_{res}(\Delta_1)$. The relation between the resonant field and the muon hyperfine coupling constant A_μ is as follows:[35]

$$A_\mu = \frac{B_{res}(\Delta_1)}{M_s\left((\bar{\gamma}_\mu)^{-1} - (\bar{\gamma}_e)^{-1}\right)} \tag{3.3}$$

The muon coupling A_μ is larger compared with the methylene proton coupling by the ratio of the muon and the proton magnetic moments (μ_p/μ_μ):

$$A'_\mu = \frac{\mu_p}{\mu_\mu} A_\mu = \frac{1}{3.1833} A_\mu \tag{3.4}$$

and on top of that by a further factor of ca. 1.2. The latter is called the intrinsic isotope effect and ascribed to the C–Mu bond length which, owing to the anharmonicity of the potential, is larger than that of the C–H by ca. 5% in the vibrational average.[84] A'_μ is called the reduced muon coupling constant. Δ_1 resonances depend only on muon hyperfine parameters and are observed solely under anisotropic conditions where they are the most intense. The mere presence of this type of resonance is thus a most sensitive indicator for anisotropic conditions and its shape and width for the nature of the averaging motion. The critical time scale for the averaging process is given by the inverse of the hyperfine anisotropy, for axial systems $|2\pi D_\perp|$, where $D_\perp$ is the perpendicular component of the anisotropy in megahertz.

Oscillation of the spin polarization between the muon and one of the protons leads to muon-proton spin flip-flop transitions. The resonance arises from the isotropic part of the hyperfine interaction and is called Δ_0 resonance ($\Delta m_\mu + \Delta m_p = 0$). It is detectable for radicals in the solid as well as in the liquid and gaseous phase. The proton coupling constant A_p is calculated from the resonant field $B_{res}(\Delta_0)$ and the muon coupling constant A_μ with the following equation:

$$A_p = \frac{(M_s)^{-1} B_{res}(\Delta_0) \, \bar{\gamma}_e \, (\bar{\gamma}_\mu - \bar{\gamma}_p) - A_\mu \, (\bar{\gamma}_e - \bar{\gamma}_\mu + \bar{\gamma}_p)}{\bar{\gamma}_p - \bar{\gamma}_e - \bar{\gamma}_\mu} \tag{3.5}$$

For isotropic media A_μ and A_p are the isotropic muon and proton hyperfine interaction constants in megahertz (M_s: spin, γ: gyromagnetic ratio (values see below); μ: index muon; e: index electron; p: index proton). In anisotropic media A_μ and A_p are the effective hyperfine interactions which depend on the orientation of the hyperfine tensor in the external field.

The dipolar part of the Hamiltonian gives also rise to Δ_2 resonances corresponding to muon-proton spin flip-flip transitions. As these are narrow and of low intensity they are of no practical importance for disordered systems.

For liquids and static single-crystalline systems, the resonances are of Lorentzian shape. For powders there is a distribution of resonance fields due to the orientation dependence of the effective hyperfine interaction giving rise to typical powder patterns.

Gyromagnetic ratios

electron	$\bar{\gamma}_e$	2.8024×10^{10}	$T^{-1}s^{-1}$
muon	$\bar{\gamma}_\mu$	1.3550×10^{8}	$T^{-1}s^{-1}$
proton	$\bar{\gamma}_p$	4.2580×10^{7}	$T^{-1}s^{-1}$
deuteron	$\bar{\gamma}_d$	6.5349×10^{6}	$T^{-1}s^{-1}$

$\bar{\gamma} = \gamma/(2\pi)$

The calculation of the oscillation frequency in the on-resonance case is performed with the following equation:

$$\nu_{res} = \frac{A_\mu^{iso} \; A_n^{iso}}{2\bar{\gamma}_e \; B_{res}} \tag{3.6}$$

The spin population ρ_s giving the population probability of the s-orbital is derived from the exprimental (A_n^{iso}) and the tabulated $(a_{o,n})$ isotropic hyperfine coupling constants as follows (for a_o see reference 85):

$$\rho_{s,n} = \frac{A_n^{iso}}{a_{o,n}} \tag{3.7}$$

Fitting of ALC Line Shapes

Using the MINUIT χ^2 minimization program[86] the resonances were fitted with a Lorentzian or with an axial powder line shape function. The latter was described in reference 87. To be able to fit the double resonance in the spectra of the pyridinium salts around 2.8 T, which is a superposition of a Δ_0 and a Δ_1 line, a fitting routine containing both, a Lorentzian and a powder line shape function, was developed. In addition in an optimized fitting routine the center of the spectrum was now free to be chosen and no longer determined automatically. Thus it was possible to keep the center out of a resonance to avoid an influence on the fitting results. In case of the Lorentzian function the fitting routine yields the resonance field, B_{res}, the half line width at half height, HWHM, and the line intensity, dip. If a powder line shape is fitted the program gives as well the resonance position and the line intensity, but instead of HWHM the asymmetry, As, is determined.

Δ_1 resonances disappear with vanishing anisotropy. In the absence of relaxation processes and for full radical yield the limiting anisotropy in an axial system for a detectable signal corresponds to 0.1 MHz (0.37 mT HWHM). This is remarkably small compared to values of 10-30 MHz (37-110 mT HWHM) expected for static systems. The full intensity of the resonance is reached at 2 MHz (7.4 mT HWHM). Then the line becomes asymmetric and broadens without loss in intensity. Consequently, even dynamically strongly reduced values should still lead to observable signals. Δ_1 resonances also disappear due to dynamic line broadening, $i.e.$, the correlation time of the motional process corresponds to the critical time scale of $\tau_{ALC} \approx 50$ ns.

When the cyclohexadienyl radical is assumed to perform fast uniaxial rotation about an axis perpendicular to the molecular plane, the calculated parallel axial component $D_\parallel$ ($\equiv D_{zz}$ in the analysis) amounts to -6.8 MHz, $i.e.$, $D_\perp = +3.4$ MHz. For uniaxial rotation about the long axis of the radical $D_\parallel$ is +5.8 MHz. Since this will reverse the high- and low-field sides and change the width of the Δ_1 resonance, it is possible to determine the preferred rotational axis from the line shape of the powder spectra.[49]

Simulation of ALC Spectra

The simulation of ALC-μSR spectra needs the input of the muon hyperfine coupling constant and the respective proton coupling constant to calculate the corresponding spectrum. As the coupling constants are determined from the observed resonance field, this program is only able to give additional information about the line intensity and the line width of Δ_0 resonances. The values are calculated analytically and the corresponding equations are given in reference 32.

3.2 Nuclear Magnetic Resonance

3.2.1 ^{2}H NMR Technique

All ^{2}H NMR experiments were recorded on a Bruker CXP 300 spectrometer at 46.07 MHz interfaced to a Tecmag spectrometer control system. The temperature dependent spectra were obtained using the quadrupole echo sequence

$$(\pi/2)_x - \tau_1 - (\pi/2)_y - \tau_2$$

with $\pi/2$ pulses of 2.2 μs and a pulse spacing of $\tau_1 = \tau_2 = 20$ μs. A modified inversion recovery sequence

$$\pi - \tau - (\pi/2)_x - \tau_1 - (\pi/2)_y - \tau_2$$

was used to determine spin-lattice relaxation times T_{1Z}, using the quadrupole echo sequence for signal detection and varying the interval τ. Instead of the inversion π pulse a composite pulse was used in these experiments which is given by $(\pi/2)_\phi(\pi/2)_{(\phi\pm(\pi/2))}(\pi/2)_\phi$ with phase cycling ($\phi = 0, \pi/2, \pi, 3\pi/2$).[88] Recycle delays took at least ten times the spin-lattice relaxation time T_{1Z}. Experimental spin-lattice relaxation times T_{1Z} were determined by analysing the amplitudes of the experimental free induction decay (FID). 16 or 32 scans were averaged for one spectrum. The sample temperature was controlled with a Bruker BVT 100 control unit, and in general it was stable to within 1 K.

3.2.2 Simulations

The behavior of an $I = 1$ spin system during quadrupole echo and inversion recovery experiments is described theoretically with the help of appropriate FORTRAN programs.[89] The simulation programs are very general and account for various types of molecular motions. A numerical diagonalization of the corresponding relaxation matrices using standard software packages gives the simulated line shapes and relaxation times.[90] Data processing and simulation of the experiments were done on SUN workstations and personal computers using the NMR1 and Sybyl/Triad software packages (Tripos, St. Louis, MO).

Theoretical Background

The theoretical background of the simulations is outlined in the following.[89] For details see references 30, 53–56, 91–94. For the analysis of the quadrupole echo experiments it is assumed that the free induction decay (FID) is given by

$$S_{qe}(t, \tau_1, \tau_2) = exp(At) \; exp(A\tau_1) \; exp(A\tau_2) \; \sigma(0) \qquad (3.8)$$

with $\sigma(0)$ referring to transverse magnetization at the beginning of the experiments. $\sigma(0)$ is given by the fractional population of the N exchanging sites. A, a complex matrix of size N, is described as

$$A = i\Omega + K \qquad (3.9)$$

revealing two different parts. The elements Ω_{ii} of the diagonal matrix Ω, the imaginary part of A, describe the frequencies of the exchanging sites, whereas the real part is related to a kinetic matrix K. The nondiagonal elements of matrix K, k_{ij}, give the jump rates from site i to j; the diagonal elements k_{ii} show the sum of the jump rates for leaving site i. The elements k_{ii} also contain the residual line width in terms of $1/T_2$ reflecting contributions from homo- and heteronuclear dipolar interactions of the spin Hamiltonian. Several internal and intermolecular processes can be superimposed depending on the complexity of the system under investigation. It is assumed that the various types of motion are independent of each other. Thus, cross terms between the different processes are neglected. Other motional models including jump motions and diffusive processes are explained in detail in references 55, 91. Equation 3.8 is solved numerically using standard diagonalization routines[90] and thus leads to ^{2}H NMR line shapes, partially relaxed spectra, and spin-spin relaxation times.

Inversion recovery experiments yield partially relaxed ^{2}H NMR spectra which are simulated using the following expression (for the meaning of τ_1, τ_2 and τ see chapter 3.2.1):

$$S_{ir}(t, \tau_1, \tau_2, \tau) = \left[1 - 2exp\left(-\frac{\tau}{T_{1Z}}\right)\right] S_{qe}(t, \tau_1, \tau_2) \qquad (3.10)$$

Here, S_{ir} and S_{qe}, are the signals from the (modified) inversion recovery experiment and the quadrupole echo experiment in the fast exchange limit, respectively. The corresponding spin-lattice relaxation time T_{1Z} is obtained via the spectral densities J_m as given by:

$$\frac{1}{T_{1Z}} = \frac{3\pi}{4} \left(\frac{e^2qQ}{h}\right) [J_1(\omega) + 4J_2(2\omega)] \qquad (3.11)$$

with the quadrupolar coupling constant e^2qQ/h (for simulation parameters see also table 3.1). Solving the following equation leads to the spectral densities J_m for a N-site problem as described in references 61,95:

$$J_m(\omega) = 2 \sum_{a,a'=-2}^{2} d_{ma}^{(2)}(\theta') d_{ma'}^{(2)}(\theta') \sum_{n,l,j=1}^{N} X_l^{(0)} X_l^{(n)} X_j^{(0)} X_j^{(n)} \qquad (3.12)$$

$$d_{0a}^{(2)}(\theta_l'') d_{0a}^{(2)}(\theta_j'') cos(a\phi_l - a'\phi_j) \frac{\lambda_n}{\lambda_n^2 + \omega^2}$$

with $\phi_i = \phi_i'' - \phi'$.

$X^{(n)}$ and λ_n are the corresponding eigenvectors and eigenvalues of the symmetrized rate matrix K'. θ'' and ϕ'' are the polar angles between the principal axis and the intermediate axis (motional axis) system, and θ' and ϕ' are those between the intermediate and the laboratory axis system, respectively. $d_{ab}^{(2)}(\theta)$ represent the elements of the reduced Wigner rotation matrix. If there is a superposition of several motional modes, then the transformation from the principal axis system to the laboratory frame is subdivided into several steps according to the number of motional contributions. Equation 3.12 describes the most general case to calculate the spin-lattice relaxation time T_{1Z}. For a series of simple molecular processes, like 3-fold or 6-fold jumps or free rotation, analytical expressions have been developed.[61,89,95] They are easily implemented into the simulation program for inversion recovery experiments (3.10).

The correlation times τ_c obey the detailed balance and are defined by:

$$\frac{1}{\tau_c} = \frac{k_{ij}}{p_j} \qquad (3.13)$$

Here, k_{ij} and p_j are the rate constant for the jump from site i to j and the equilibrium population of site j, respectively.

Correlation of Simulation and Experiment

The correlation of the simulations and the experimental data acquisition is displayed in figure 3.2. Quadrupole echo experiments on the sample give the experimental line shapes, which can be simulated by input of population values or a jump angle, depending on the type of simulation used. Population and jump angle then can be used to determine thermodynamic data. Inversion recovery experiments with given delay times yield the experimental partially relaxed spectra, which can be mimicked by the input of correlation times. The latter enable the calculation of kinetic data. Both, the experimental partially relaxed line shapes as well as the simulated ones give the spin-lattice relaxation time T_{1Z} and consequently the experimental and the simulated activation energy.

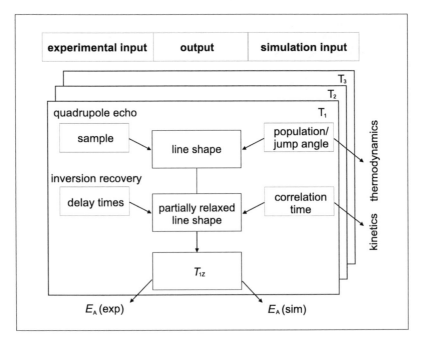

Figure 3.2: Correlation of simulations and experimental data acquisition showing input and output data as well as deduced variables.

Applied Simulation Programs

Simulations of the quadrupole echo and the inversion recovery experiments were
performed with a 2-site, a 3-site and a 6-site jump model being described in detail
below.

2-site Jump Model. In the 2-site jump model a jump process between two different equally populated ($p_1 = p_2 = 0.5$) orientations is simulated. Adjustment of
the simulated to the experimental line shape is achieved varying the jump angle
between $0°$ and $90°$. A $90°$ jump is synonymous to a fast free rotation.

3-site Jump Model. In this study a 3-site jump model was used which mimics
in plane jump processes of the perdeuterated pyridinium cations between three
different orientations, synonymous with $120°$ jumps and consistent with the $R\bar{3}m$
symmetry of the high temperature phase. If the jump process is independent of
its history it is a Markov process. The rotational axis is illustrated in figure 2.4.
The population probabilities (in the following: populations) p_1, p_2 and p'_2, with

$p_2 = p_2'$ and $p_1 + p_2 + p_2' = 1$, correspond to the three orientations. The ratio of the populations gives the equilibrium constant K for an individual jump process

$$0° \quad \rightleftharpoons \quad \pm 120° \tag{3.14}$$

$$p_1 \quad \overset{K_{12}}{\rightleftharpoons} \quad p_2 = p_2' \tag{3.15}$$

$$K_{12} = \frac{p_2}{p_1} = \frac{p_2'}{p_1} = \frac{1 - p_1}{2p_1} \tag{3.16}$$

and therefore the free enthalpy $\Delta G° = -RT \ln K_{12}$ describing the energy offset between the corresponding orientations (see figure 3.3). The overall potential V, which is formaly a superposition of a 3-fold (V_3) and a 1-fold potential (V_1), is given by:

$$V(\text{3-site}) = V_1(\Psi) + V_3(\Psi) \tag{3.17}$$

$$= \frac{V_1°}{2} \sin\left(\Psi - \frac{\pi}{2}\right) + \frac{V_3°}{2} \sin\left(3\Psi - \frac{\pi}{2}\right)$$

$$\text{with:} \qquad \Delta G° = \frac{3}{4} V_1° \tag{3.18}$$

6-site Jump Model. In addition a 6-site jump model was used which mimics in plane jump processes between six different orientations, synonymous with 60° jumps. This simulation model takes into account the 3-fold rotation-inversion axis found in the high temperature phase. Furthermore it is plausible based on the pseudo 6-fold symmetry of the pyridinium ion. Like in the 3-site jump model the jumps are regarded as Markov processes. The rotational axis is identical to that in the 3-site jump model (see figure 2.4). The population probabilities (in the following: populations) p_1 (0°), p_2 and p_2' ($\pm 60°$), p_3 and p_3' ($\pm 120°$) and p_4 (180°) with $p_2 = p_2'$, $p_3 = p_3'$ and $p_1 + p_2 + p_2' + p_3 + p_3' + p_4 = 1$, correspond to the six orientations.

$$0° \quad \rightleftharpoons \quad \pm 60° \quad \rightleftharpoons \quad \pm 120° \quad \rightleftharpoons \quad 180° \tag{3.19}$$

$$p_1 \quad \overset{K_{12}}{\rightleftharpoons} \quad p_2 = p_2' \quad \overset{K_{23}}{\rightleftharpoons} \quad p_3 = p_3' \quad \overset{K_{34}}{\rightleftharpoons} \quad p_4 \tag{3.20}$$

The 6-site model consists of a 6-fold potential V_6 superimposed by a 1-fold potential V_1 and is described as follows:

$$V(\text{6-site}) \;=\; V_1(\Psi) + V_6(\Psi) \tag{3.21}$$

$$=\; \frac{V_1^\circ}{2}\, sin\left(\Psi - \frac{\pi}{2}\right) + \frac{V_6^\circ}{2}\, sin\left(6\Psi - \frac{\pi}{2}\right)$$

with:
$$\Delta G^\circ \;=\; \frac{1}{4}\, V_1^\circ \tag{3.22}$$

As a consequence of the superposition of the two potentials the relation of the population probabilities is given by:

$$p_1 \;=\; \frac{1}{1 + 2exp(a) + 2exp(3a) + exp(4a)} \tag{3.23}$$

$$\begin{aligned}
p_2 &\;=\; p_1\, exp(a) = p_2' \\
p_3 &\;=\; p_1\, exp(3a) = p_3' \\
p_4 &\;=\; p_1\, exp(4a)
\end{aligned}$$

$$a = -\frac{\Delta G^\circ}{RT} \tag{3.24}$$

The free energy offset between orientations p_1 and p_2 as well as between p_3 and p_4 amounts to ΔG°, whereas the energy difference between orientation p_2 and p_3 amounts to the double. The ratio of the populations gives once more the equilibrium constant of the idividual jump process:

$$K_{12} = \frac{p_2}{p_1} \qquad\qquad K_{23} = \frac{p_3}{p_2} \qquad\qquad K_{34} = \frac{p_4}{p_3} \tag{3.25}$$

The equilibrium constant K for the whole circuit is given as the product of the individual equilibrium constants:

$$\begin{aligned}
K &\;=\; K_{12} \cdot K_{23} \cdot K_{34} \cdot K_{43} \cdot K_{32} \cdot K_{21} \\
&\;=\; K_{12} \cdot K_{23} \cdot K_{34} \cdot \frac{1}{K_{34}} \cdot \frac{1}{K_{23}} \cdot \frac{1}{K_{12}} = 1
\end{aligned} \tag{3.26}$$

For both models, the 3-site as well as the 6-site jump model, the quadrupole echo line shapes were simulated by solely varying the population p_1 at a fixed value of the correlation time τ_c in the fast exchange limit. The assumption of fast exchange limit spectra was verified by T_2 relaxation experiments. The additional population values (3-site: p_2; 6-site: p_3 and p_4) are related to p_1 by the 1-fold potential V_1° as described above.

Inversion recovery spectra were optimized by only modifying the correlation time τ_c at fixed population values that were determined from equilibrium spectra. The final best fit was chosen by comparison of the superimposed experimental and theoretical spectra, not by modulation of the simulated T_{1Z} value, taking into account the overall line shapes as well as the relative amplitudes in case of the partially relaxed spectra. The potential curves corresponding to the 3-site and the 6-site jump model are displayed in figure 3.3. Complementing simulation parameters are given in table 3.1.

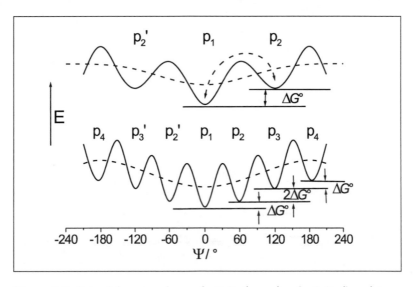

Figure 3.3: Potential energy scheme of a 3-site (upper) and a 6-site (lower) jump model.

Table 3.1: Simulation parameters used during the ^{2}H NMR data analysis (quadrupole echo, inversion recovery experiments).

Parameter	Value
quadrupole echo experiments: population p_1	variable (see fig. 6.2)
inversion recovery experiments: correlation time τ_c	variable (see fig. 6.7)
quadrupolar coupling constant ^{a} (e^2qQ/h)	185 kHz
transformation angles ^{b} (deg) ψ (3-site) ψ (6-site)	-120; 0; +120 -180; -120; -60; 0; +60; +120; +180
residual line width $(1/(\pi T_2))$	1000-4000 Hz

^{a}Asymmetry parameter $\eta=0$. ^{b}Euler angles $\phi=0°$, $\theta=90°$, ψ relating the magnetic principal axis and the molecular axis system (z-axis parallel to motional axis).

3.3   Differential Scanning Calorimetry

Heating rate dependent DSC measurements were performed on a NETZSCH DSC-204 for the pyridinium tetrafluoroborate and the perchlorate, both compounds in one case perdeuterated (PyBF$_4$-d5, PyClO$_4$-d5) in the other case non-deuterated (PyBF$_4$, PyClO$_4$), using an aluminium sample cell against an empty reference cell of the same type. The tetrafluoroborate samples were treated according to the program (a), the perchlorate samples according to the program (b)

(a)   153 K $\xrightarrow{5\,\text{min}}$ 153 K $\xrightarrow{r}$ 273 K $\xrightarrow{5\,\text{min}}$ 273 K $\xrightarrow{r}$ 153 K

(b)   183 K $\xrightarrow{5\,\text{min}}$ 183 K $\xrightarrow{r}$ 283 K $\xrightarrow{5\,\text{min}}$ 283 K $\xrightarrow{r}$ 183 K

with r giving the heating rate of 2 K min^{-1}, 5 K min^{-1} and 10 K min^{-1}. For all samples the program cycle was repeated twice. The molar heat capacity c_p was obtained using the heat flux $\dot{Q} = dQ/dt$ per sample mass m, the heating rate $r = dT/dt$, and the molar mass $M = m/n$:

$$c_p = \frac{M}{m}\frac{\dot{Q}}{r} = \frac{1}{n}\frac{dQ}{dT} \qquad (3.27)$$

3.4   Sample Preparation

Water employed as a solvent was always distilled twice before use.

3.4.1   Zeolites

Preparation of NaY and NaX

Zeolite NaY was kindly provided by CU Chemie Uetikon, Uetikon, Switzerland. Zeolite NaX was purchased from Aldrich Chemicals. Both, NaY and NaX, were prepared always in the same manner before use for metal exchange.

To avoid template residues the zeolite was calcined under atmosphere according to the following temperature program:

293 K $\xrightarrow{0.5\,\text{K min}^{-1}}$ 783 K $\xrightarrow{10\,\text{h}}$ 783 K $\xrightarrow{}$ 293 K

After stirring the zeolite in an aqueous sodium chloride solution (molar ratio: n(zeolite)/n(NaCl) $\approx$ 1/45; c(NaCl in H$_2$O) $\approx$ 0.1 mol l^{-1}) for about 20 h at room temperature, it was filtered, washed with water and dried at room temperature. The zeolite was then calcined again and stored above a saturated CaCl$_2$ solution to ensure a well defined and constant water content of 22 wt%.

Metal Exchange

All metal exchange procedures were done via aqueous ion exchange using the corresponding appropriate metal salts all purchased from Aldrich Chemicals except the silver salt which was from Merck. The minimum amount of metal salt was calculated taking into account the number of exchangeable sodium cations as well as the charge ratio of Na^+ and the corresponding metal ion. In zeolite NaY only 30 of the 58 sodium cations in one unit cell are accessible via a single step aqueous ion exchange to be replaced by another metal cation; in zeolite NaX it is 56 of 80 Na^+ in a unit cell (see also chapter 2.4.1). In NaY all exchangeable sodium ions are located on site II, *i.e.*, in the supercage. Thus in the following we only refer to the supercage for simplification, not to the unit cell. As one unit cell contains eight supercages and 30 exchangeable sodium ions 3.75 positive charges per supercage can be replaced in maximum.

All ion exchange procedures were performed with a metal ion excess with respect to the amount of exchangeable sodium ions to ensure complete exchange of the accessible sodium ions in the supercage. This means the amount of the silver salt was calculated to achieve an exchange of four metal ions per supercage. In case of the +2 charged metal ions the amount of metal salt was calculated to achieve two metal ions per supercage. Only in the case of manganese there was an aim at 3.5 ions per supercage in NaX, not in NaY, as a crystal structure determination for a much higher loading in this zeolite exists.[96]

The water content of all zeolite samples after metal exchange was determined by atom emission spectroscopy/induced coupled plasma (AES/ICP) to be able to calculate the metal content of the samples. Only the platinum, the nickel, the zinc and the manganese exchanged zeolites were then accessible by x-ray fluorescence analysis (RFA) to acquire the amount of exchanged metal cations. If the experimentally determined amount of sodium ions plus the charges from the exchanged metal ions did not suffice to reach charge balance with the zeolite framework, protons were added to achieve a neutral chemical formula (see below).

PtNaY $[Pt(NH_3)_4]\,Cl_2 \cdot xH_2O$ (99.99%)   was   dissolved   in   water ($c_{Pt} = 0.023$ mol l^{-1}) and added dropwise to a suspension of the zeolite in water ($c_{NaY} = 10$ g l^{-1}) over 4 h. The mixture ($n(Pt^{2+})/n(Na^+) = 0.4^1$; $Pt^{2+}/sc = 2$) was stirred for 48 h at 338 K. After filtering the zeolite was washed with water and dried under vacuum at room temperature. To release the amine ligands calcination was performed according the following temperature program in an oxygen

---

[1]The ratio of platinum and sodium ions corresponds to the total amount of Na^+ in the unit cell. Taking into account that only 52% of Na^+ is accessible for ion exchange in NaY the ratio is enlarged to $n(Pt^{2+})/n(Na^+_{access}) = 0.77$. The fact that the charge of a platinum cation is only compensated if two sodium ions are released enhances the ratio once again to $n(Pt^{2+})/n(2Na^+_{access}) = 1.5$. Consequently platinum was used in 1.5 fold excess in the described ion exchange procedure with respect to the amount of exchangeable sodium ions.

flow of 235 ml min^{-1}:

$$293 \text{ K} \xrightarrow{0.5 \text{ K min}^{-1}} 523 \text{ K} \xrightarrow{4 \text{ h}} 523 \text{ K} \xrightarrow{2 \text{ K min}^{-1}} 293 \text{ K}$$

The metal exchanged zeolite was stored above a saturated $CaCl_2$ solution and the water content was determined by EA to be 17%. In addition the elementary analysis revealed that the sample still contains huge amounts of nitrogen indicating that calcination did not work properly. RFA analysis observed a platinum content of 0.75 platinum ions per supercage leading to a chemical formula $H_{18}Na_{20}Pt_6Si_{134}Al_{58}O_{384} \cdot 30NH_3$.

PdNaY The preparation of PdNaY was performed analogous to the preparation of the platinum exchanged zeolite but starting from $[Pd(NH_3)_4]Cl_2 \cdot xH_2O$ (99.99+%). The maximum temperature during calcination was 773 K. PdNaY was not accessible to RFA.

NiNaY $NiCl_2 \cdot 6H_2O$ (99.9999%) was dissolved in water ($c_{Ni} = 0.088$ mol l^{-1}) and added dropwise to a suspension of the zeolite in water ($c_{NaY} = 15$ g l^{-1}). The mixture ($n(Ni^{2+})/n(Na^+)) = 1.0$; $Ni^{2+}/sc = 2$) was stirred 24 h at room temperature. After filtering the zeolite was washed with water, dried under vacuum at room temperature and stored above a saturated $CaCl_2$ solution. The nickel content of NiNaY determined by RFA amounts to 1.4 Ni^{2+}/sc.

AgNaY The preparation of AgNaY was performed in a darkroom according to the preparation of the nickel exchanged zeolite starting from $Ag(NO_3)$ (99.8-100.5%). The water content of the metal exchanged zeolite was 18%. AgNaY was not accessible to RFA.

ZnNaY The preparation of ZnNaY was performed according to the preparation of the nickel exchanged zeolite but starting from $Zn(NO_3)_2 \cdot xH_2O$ ($x \approx 6$; 99.999%). The water content of the metal exchanged zeolite was determined to be 23%. The RFA analysis revealed a zinc amount of 1.9 ions per supercage. This gives a chemical formula $Na_{28}Zn_{15}Si_{134}Al_{58}O_{384}$.

MnNaX The preparation of MnNaX was performed according to the preparation of the nickel exchanged zeolite but starting from $MnCl_2 \cdot 4H_2O$ (99.99%). Before using the solvent it was bubbled with nitrogen in an ultrasonic bath for 24 h and the reaction also took place under nitrogen atmosphere. The water content of the metal exchanged zeolite amounted to 18%. RFA analysis gave a metal content of 2.25 manganese ions per supercage leading to a chemical formula $H_8Na_{36}Mn_{18}Si_{112}Al_{80}O_{384}$.

Benzene Loading

The minimum amount of benzene was calculated assuming that all accessible sodium ions were replaced by the corresponding metal cations (this is 15 metal cations charged +2 or 30 metal cations charged +1 in NaY, and 28 metal cations charged +2 in NaX) and that a single benzene molecule will interact with a single metal cation. In addition the total amount of benzene should not exceed three molecules per supercage to ensure uniform distribution and to avoid benzene aggregates. Taking into account that the metal exchange was probably not complete all experiments were performed with an deficit amount of benzene to avoid benzene molecules on different adsorption sites; that is approximately one benzene per supercage in case of the +2 charged metal ions and two benzene molecules in case of silver (see also table 3.2).

A stainless steel cell (see fig. 3.4) was filled with metal exchanged, hydrated zeolite and sealed by welding. This worked well for all samples except the ZnNaY. The cell for the latter was therefore screwed using a WILLS gasket - a hollow, silver-plated copper gasket filled with nitrogen gas - for sealing. All samples were then dehydrated under vacuum according to the following temperature program:

$$333 \text{ K} \quad \longrightarrow \quad 373 \text{ K} \quad \longrightarrow \quad 433 \text{ K} \quad \longrightarrow \quad 473 \text{ K}$$

The different temperatures were kept for at least 12 h. The benzene was purified from oxygen via freeze-pump-thaw cycling at least three times, then sublimed and adsorbed on the zeolite. After loading with benzene the sample cells were directly sealed with a cold welding crimping tool. All cells were kept at 353 K for at least 24 h to achieve equlibrium.

After measuring the samples with μSR, the cells were opened to investigate the benzene content by elementary analysis (EA). For the results of the metal exchanged zeolites see table 3.2. In case of platinum, zinc and manganese the amount of benzene molecules per supercage (benz/sc) was calculated with respect to the chemical formula determined via RFA (see also chapter 3.4.1). In case of palladium, nickel and silver the benzene content was determined referring to the maximum amount of benzene, $(\text{benz/sc})_{max}$, given also in table 3.2. Elementary analysis for the pure zeolite NaY observed 4 molecules per supercage which is 90% of the calculated value. The pure zeolite NaX was only loaded with 0.4 molecules per supercage. That is 8% of the calculated maximum value.

3.4.2   Pyridinium Salts

Non-deuterated Salts

Both non-deuterated pyridinium salts, **PyBF$_4$** and **PyClO$_4$**, were provided by Dr. P. Czarnecki and coworkers. The preparation of the samples was done according to reference 10. ALC-μSR measurements on the pyridinium salts were performed in homemade brass cells sealed with an indium gasket (see fig. 3.4). Transverse field measurements were carried out in spherical DURAN glass bulbs of 25 mm diameter.

Perdeuterated Salts

Pyridine-d$_5$ (99 atom% D) and tetrafluoroboric acid (48 wt.% solution in water) were purchased from Aldrich Chemicals and used without further purification. Pyridine-d5 was added to the tetrafluoroboric acid solution at room temperature under stirring without using an additional solvent (molar ratio: n(Py-d$_5$)/n(HBF$_4$)=1). **PyBF$_4$-d$_5$** crystals formed immediately and were measured after drying without further purification.

The preparation of **PyClO$_4$-d$_5$** was performed according to the same procedure using perchloric acid (60 wt.% solution in water; molar ratio: n(Py-d$_5$)/n(HClO$_4$)=1) which was also purchased from Aldrich Chemicals.

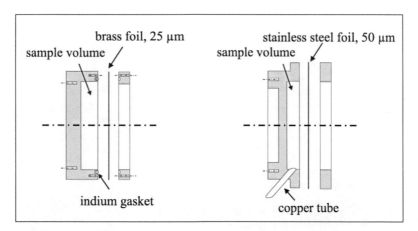

Figure 3.4: Homemade sample cells for ALC-μSR measurements (left: brass cell; right: stainless steel cell).

Table 3.2: Preparation parameters of metal exchanged zeolite samples loaded with benzene.

	NaY			NaX				
	Pt	Pd	Ni	Ag	Zn	Mn	units	method
c_{zeol}	10	10	15	15	15	15	g l^{-1}	
M_{salt}	334.1	245.5	237.7	169.9	297.5	197.9	g mol^{-1}	
c_{salt}	0.023	0.027	0.088	0.14	0.087	0.10	mol l^{-1}	
$\left(\dfrac{Me^{n+}}{Na^+}\right)_{max}$	0.40	0.46	1.0	1.6	1.0	0.83		calc.
$\left(\dfrac{Me^{n+}}{sc}\right)_{max}$	2.0	2.0	2.0	4.0	2.0	3.5		calc.
$\left(\dfrac{benz}{Me^{n+}}\right)_{max}$	0.50	0.50	0.60	0.50	0.70	0.37		calc.
$\left(\dfrac{benz}{sc}\right)_{max}$	0.90	0.90	1.1	2.1	1.4	1.3		calc.

zeol = zeolite; salt = metal salt; Me^{n+} = +n charged metal ion; sc = supercage
$M(NaY) = 17116$ g mol^{-1}; $M(NaX) = 17780$ g mol^{-1}

	Pt	Pd	Ni	Ag	Zn	Mn	units	method
H_2O	17	-^{a}	-^{a}	18	23	18	wt.%	AES
$\left(\dfrac{Me^{n+}}{sc}\right)_{exp}$	0.75	-^{a}	1.4	-^{a}	1.9	2.3		RFA
$\left(\dfrac{benz}{Me^{n+}}\right)_{exp}$	1.2	-	0.79	-	0.79	0.52		
$\left(\dfrac{benz}{sc}\right)_{exp}$	0.90^b	0.10^c	1.1	1.9	1.5	1.2		EA

AES = AES/ICP: atom emission spectroscopy/induced coupled plasma; RFA: x-ray fluorescence analysis; EA: elementary analysis.
^{a}: not determined; ^{b}: from theoretical values; ^{c}: value for the ocher colored zeolite material.

Chapter 4

μSR on BENZENE in FAUJASITE

4.1   NaY

4.1.1   Background

Benzene in zeolite NaY has been well studied by ^{2}H NMR,[68] neutron scattering[65] and molecular dynamics studies.[97]   NMR experiments on benzene in faujasite revealed that the π-system interacts with sodium ions in the supercage.   The molecule maintains a high rotational mobility about the 6-fold axis.[72] And only recently, Fleming and coworkers reported about hyperfine and host-guest interactions of the muonated cyclohexadienyl radical, generated via the muonium addition to benzene, in NaY. In concert with studies of the parent benzene molecule, as well as current theoretical calculations, the dominant adsorption site of the C_6H_6Mu radical is believed to be the SII Na cation, within a supercage next to a 6-ring window, which gives rise to two observed Δ_1 ALC lines. Those lines correspond to two different orientations for the muon of the CHMu methylene group pointing towards (endo) and away (exo) from the sodium cation. The cation interaction gives rise to unprecedentedly large (≈20%) shifts in hyperfine coupling constants, indicative of a strong bond formed with the π electrons of the C_6H_6Mu radical. An additional but weaker Δ_1 resonance line is interpreted as being due to adsorption at the window site (12-ring window between the supercages). The ALC lines associated with the C_6H_6Mu radical bound to both the Na cation and the window sites are all broad, ≈0.1 T, change little with temperature and exhibit mainly static line shapes over the whole temperature range studied, from 3 K to 322 K. This indicates a much stronger host-guest interaction for C_6H_6Mu, particularly with the Na cation, than is known for benzene, to the extent that this site acts as an effective trap for the free radical, over the critical μSR time scale of 50 ns.[39]

Table 4.1: Literature data for hyperfine coupling constants (in MHz) of benzene and of the muonated cyclohexadienyl radical in zeolite NaY.

system	T/K	A_μ	A'_μ	A_p	reson.	ref.
benzene, gaseous	313	507.5	159.4	124.9		98
benzene, liquid	300	514.5	161.6	126.1		32, 99
benzene, frozen	263	524.5	164.7	128.2		32, 36, 39, 84
C_6H_6Mu/NaY	322	605.2	190.1	108.1	B/A	39
C_6H_6Mu/NaY	322	529.3	166.2	≈ 130	C	39
C_6H_6Mu/NaY	322	430.0	135.1	160	D/(E)	39

4.1.2   Experimental Results and Discussion for C_6H_6/NaY

The zeolite NaY samples loaded with benzene investigated in the present context are different to those from Fleming et $al.$[39] with respect to temperature treatment. After loading with benzene the Fleming samples were allowed to equilibrate at room temperature over several days, whereas the present samples were stored at 353 K for at least 24 h. In addition the samples in this context have higher loadings (4 C_6H_6/sc $\equiv$ 15 wt.% compared to 2-3/sc). To account for this and for the sample design NaY was studied in some detail to allow a direct comparison with the metal exchanged zeolite samples and thus to identify the influence of the metal cation. C_6H_6/NaY was investigated in a temperature range from 96 K to 471 K with 10 mT step size in a field range from 1.0 T to 3.0 T. Representative spectra are displayed in figure 4.1. With respect to Fleming and coworkers the resonances are labeled A, B, C and D (downward field direction). Additional resonances found during the investigations presented here or not dubbed by Fleming et $al.$ are named E, F and G (upward field direction). All resonance positions, the corresponding line widths and in case of assignment the hyperfine coupling constants in dependence of temperature are given in tables 4.2 (for B, C, and D) and 4.3 (for A, E, and G).

Hyperfine Coupling Constants

Despite the different sample characteristics the μSR spectra of benzene loaded zeolite NaY investigated in the present context are at first sight identical to those of Fleming et $al.$ with respect to resonance position, shape and intensity.

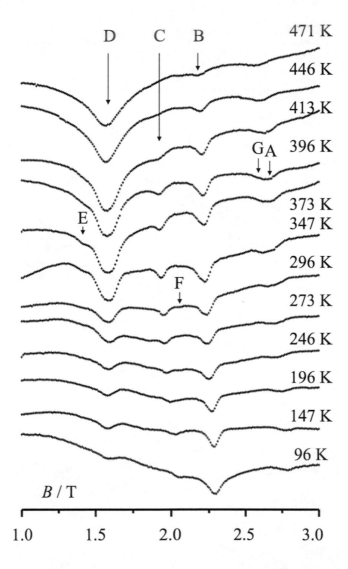

Figure 4.1: Temperature dependent ALC-μSR spectra of NaY loaded with 15 wt.% benzene (4 C_6H_6/sc). B, C and D indicate the Δ_1 resonance positions, E, F, G and A the Δ_0 resonance positions. (A, B, C, D: according to the labeling of Fleming *et al.*;[39] E, F, G: new labeling.)

Table 4.2: Resonance positions (B_{res} in T), hyperfine coupling constants (A'_μ in MHz) and Lorentzian widths W (HWHM in mT; axial line width for B: D_{zz} in MHz) of Δ_1 resonances in benzene loaded NaY (4 molecules/sc).

	D			C			B		
T/K	B_{res}	A'_μ	W	B_{res}	A'_μ	W	B_{res}	A'_μ	D_{zz}
471	1.567	134.7	132.4	1.868	160.6	65.9	2.195	188.6	-10.61
446	1.569	134.9	116.6	1.882	161.8	91.2	2.202	189.2	-10.72
413	1.575	135.4	107.5	1.909	164.1	61.1	2.209	189.9	-10.97
403	1.578	135.7	101.3	1.912	164.4	54.2	2.211	190.0	-12.22
395	1.579	135.7	102.5	1.917	164.7	53.0	2.213	190.2	-12.49
373	1.581	135.9	92.1	1.925	165.4	36.6	2.217	190.6	-13.43
347	1.578	135.6	83.4	1.932	166.1	32.2	2.221	190.0	-13.34
297	1.580	135.8	74.5	1.949	167.6	36.0	2.237	192.3	-12.50
296	1.579	135.7	71.2	1.950	167.6	29.9	2.238	192.3	-12.48
273	1.579	135.7	74.1	1.958	168.3	27.7	2.242	192.7	-13.72
246	1.570	135.0	86.3	1.968	169.2	26.5	2.254	193.8	-12.02
196	1.562	134.3	86.9	1.995	171.5	40.4	2.272	195.3	-10.72
147	1.575	135.4	63.8	2.024	173.9	53.6	2.287	196.6	-9.65
96	1.558	133.9	78.6	2.048	176.0	40.9	2.294	197.2	-5.77

Table 4.3: Resonance positions (B_{res} in T), hyperfine coupling constants (A_p in MHz) and Lorentzian widths W (HWHM in mT) of Δ_0 resonances in benzene loaded NaY (4 molecules/sc).

	E			G			A		
T/K	B_{res}	A_p^D	W	B_{res}	W		B_{res}	A_p^B	W
403	1.401	167.4	71.9	2.544			2.657	105.6	
395	1.399	167.9	73.1	2.613	93.4		2.673	103.0	41.0
373	1.406	167.2	54.2	2.620	78.6		2.690	101.1	44.9
347	1.408	166.0	56.0	2.621	48.1		2.691	102.1	44.8
297	1.427	163.0	120.6	2.656	52.1		2.721	100.8	41.5
296	1.437	161.1	140.7	2.649	47.0		2.714	102.2	41.9
273				2.669	55.0		2.725	101.4	43.8
246				2.680	34.0		2.732	103.4	46.1
196				2.696			2.763	102.5	

Therefore the labeling for resonances A, B, C, and D was adopted. Fleming and coworkers used TF-μSR to identify the resonances B, C, and D to be Δ_1 lines; B and D corresponding to the cyclohexadienyl radical bound to sodium ions in the SII position and C situated at the window site. B and D represent two different orientations of the radical, namely with the muon in exo and in endo position with respect to the cation, respectively. This results in muon hyperfine couplings which are increased by $\approx$20% for resonance B (192.3 MHz at 296 K) and decreased for resonance D (135.7 MHz at 296 K) compared to the nearly unperturbed cyclohexadienyl radical at the window site (C: 167.6 MHz at 296 K). As the deviations in hfc of resonances B and D originate from geometrical reasons, the corresponding methylene proton hfcs are affected in the opposite direction. That is why resonance A was identified to be related to the B line, i.e., the Δ_0 resonance of the endo methylene proton with a hyperfine coupling constant that is by $\approx$20% decreased (102.2 MHz at 296 K). For the Δ_1 resonances the resonance positions of Fleming and coworkers and in this work differ in average only by about $\pm$0.008 T with a maximum deviation of 0.02 T. Only for the Δ_0 resonance A the differences are more obvious. That ist about $\pm$0.03 T with a maximum difference of 0.04 T. This is not surprising as resonance A is involved in a double resonance, which was not resolved by Fleming et al. but in this work. The hyperfine coupling constants for resonances B, C, D, and A were calculated and are displayed in figure 4.2 (solid symbols). To enable direct comparison the values of Fleming and coworkers are also shown in the plots.

As already obvious from the resonance positions the temperature dependences of the hfcs are in very good correspondence with the Fleming data for B, C, and D. For resonance B it amounts to -0.024$\pm$0.001 (dA'_μ/dt given in MHz K^{-1}), for resonance C to -0.038$\pm$0.001. That is identical within error to the values of Fleming et al.. Yu et al. obtained a temperature dependence of the muon hyperfine coupling in pure liquid benzene of -0.025 MHz K^{-1}.[100] Fleming did not observe a temperature dependence of the hfc for resonance D, and neither do we. The temperature dependence of the hfc for resonance A (dA_p/dt) was calculated to be +0.006$\pm$0.007. Already the error shows that the determination of a slope is difficult. The value differs obviously to that found by Fleming et al.; but in contrast to the Δ_1 resonances it is clearly positive. It is important to point out that the trend of the hfcs is even followed at higher temperatures than measured by Fleming et al. Consequently the higher loading does not influence the resonance positions.

Compared to an unperturbed cyclohexadienyl radical the hyperfine coupling constant of the Δ_1 resonance B is scaled up. In contrast, the corresponding proton Δ_0 resonance A is scaled down. The corresponding proton Δ_0 line to resonance D is expected to behave in the same way. As resonance D reveals a decrease in the hfc, the corresponding Δ_0 line is expected at a hfc scaled up to 160$\pm$6 MHz related to a resonant field of approximately 1.4 T. In fact a very weak resonance

E at the low field branch of resonance D was observed in the present context. But in contrast to Fleming who only had a hint on this line it was possible to fit this signal over a temperature range from 296 K to 403 K. The temperature dependence of the hfc (161.1 MHz at 296 K) amounts to +0.056±0.009 MHz K^{-1}. As for the other Δ_0 line it is positive, but much larger. The temperature dependence of the methylene proton hyperfine coupling in pure liquid benzene was determined by Yu *et al.* to amount to -0.012 MHz K^{-1}.[100]

The Δ_0 line corresponding to resonance C is expected to occur around a field value of the unperturbed cyclohexadienyl radical, *i.e.*, at 2.15 T, related to a hfc of ≈130 MHz. In this work only at 296 K a resonance F was detected at 2.0538 T giving a hfc of 146.9 MHz if regarded as Δ_0 line of the C resonance. Although the hfc is larger than predicted, it is most likely that resonance F is related to resonance C.

An additional resonance G was detected over the whole temperature range being involved in a double resonance with line A. Assigning this line as a proton Δ_0 resonance of one of the Δ_1 resonances yields rather implausible hfcs and is therefore excluded. Up to now the origin of this resonance is not clear. This is rather surprising as the resonance is quite intense compared to resonances E and F. Fleming *et al.* proposed that resonance G could be a sodium Δ_0 line. We exclude this assignment as the resonance disappears upon deuteration of the cyclohexadienyl radicals and therefore has to origin from a proton interaction. Fleming and coworkers did not assign any other resonance than G to a sodium/radical interaction; and neither do we. However, it has to be taken into account that resonances B and D are broad and intense. This means it is quite likely that additional lines are superimposed by the latter. That is why we decided to investigate NaY loaded with perdeuterated benzene as well. The experimental results and the analysis are given in chapter 4.1.3.

The temperature dependence of the hfcs are displayed in figure 4.2, the data are given in tables 4.2 and 4.3.

Line Shape

As resonance B shows an axial line shape it was fitted with an axial powder function. To enable a comparison of the line widths the D_{zz} value obtained from the powder fit was converted into HWHM (half width at half maximum):

$$\frac{D_{zz} \text{ (MHz)}}{2\gamma_\mu (\text{MHz T}^{-1})} = \text{HWHM(T)}. \tag{4.1}$$

In contrast, resonances C and D were were fitted with a Lorentzian line. Resonance A is involved in a double resonance. Therefore it was fitted with a double function consisting of two Lorentzian lines. It has to be noted that the error for fitting resonances A, B, and C was within limit; for resonance D it was

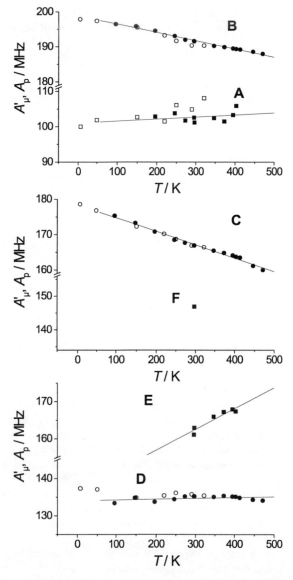

Figure 4.2: Temperature dependence of hfcs in NaY loaded with 15 wt.% benzene (4 C_6H_6/sc) (solid symbols). (Circles give the hfcs of Δ_1, squares of Δ_0 resonances. Open symbols are the results of Fleming *et al.* with 2-3 C_6H_6/sc to permit comparison.[39] Solid lines give the linear fit to the present experimental data.)

considerably larger. This could be due to the fact that this resonance at some temperature points resembles more an axial line shape, but fitting with a powder function does not give any improvement. This in fact could be a hint that resonance D is possibly a superposition of two different lines. The intensity of the resonance would argue for this interpretation, too.

Fleming and coworkers fitted all resonances with Gaussian line shapes. The given 2σ widths are converted into HWHM to enable comparison (FWHM is the full width at half maximum):

$$HWHM=0.5\cdot FWHM=0.5\cdot 2.35\cdot \sigma=0.5\cdot 1.175\cdot 2\sigma=0.5875\cdot W$$

Resonances E, F and G were all fitted using a Lorentzian line shape. But as resonance F was detected only at one temperature point no temperature dependence was obtained.

Line Width

The line widths determined from the fits of the resonances are listed in tables 4.2 and 4.3. They are shown in figure 4.3 simultaneously with the data of Fleming and coworkers to permit direct comparison. As far as possible the data were fitted linearly. It should be taken into account that the widths, being more sensitive to background substraction, are less precise than the resonance positions and therefore the hfcs obtained from the fit. (For line width see also chapter 3.1.4.)

As resonance B was fitted with a powder function giving D_{zz} values the latter had to be converted into HWHM which are given in the plot. Although the values obtained in the present context are larger (≈ 70 mT at 300 K) than observed by Fleming *et al.* (≈ 50 mT) the trend with temperature is identical down to 250 K. Only at lower temperatures the values drop. Up to 300 K the line widths of resonances C and D are comparable to those of Fleming *et al.* (C: ≈ 30 mT; D: ≈ 70 mT at 300 K), for resonance C even more than for resonance D and the temperature dependences are quite similar. For C the slope is identical to the Fleming data being negative and small. For resonance D the slopes are also small, but the present data already show a positive value which is not very reliable as the error shows ($+6\times 10^{-5}\pm 4.7\times 10^{-2}$ mT K^{-1}). Above 300 K a marked increase of line width takes place for both lines, *i.e.*, $+0.40\pm 0.02$ mT K^{-1} for resonance C and $+0.38\pm 0.02$ mT K^{-1} for resonance D. This phenomenon was not observed by Fleming *et al.* as their study did not cover this temperature range. Thus it is not clear whether this drastic increase is due to the higher loading or whether it would take place at lower loadings as well. In addition, this phenomenon was not observed for resonance B although the latter should be comparable to resonance D.

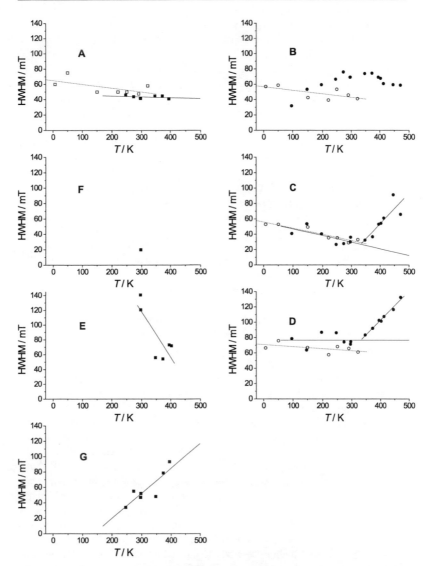

Figure 4.3: Temperature dependence of the resonance widths in NaY loaded with 15 wt.% benzene (4 C_6H_6/sc) (solid symbols). (Circles give the widths of Δ_1 (right column), squares of Δ_0 resonances (left column). Open symbols are the results of Fleming *et al.* with 2-3 C_6H_6/sc to permit comparison.[39] Solid lines give the linear fit to the present experimental data. Dashed lines are the linear fit to the data of Fleming *et al.* which were converted from Gaussian widths to HWHM.)

The line widths for resonance A are also comparable to those of Fleming *et al.* over the whole temperature range although the decrease with temperature of the present data is more modest. But the trend of the values is even followed at higher temperatures. The determination of the line width of resonance E is difficult as this line is very weak and occurs on the low field branch of the very intense resonance D. The line width of E shows a very strong decrease with increasing temperature which amounts to -0.61±0.23 mT K^{-1}. Resonance F only occured at one temperature point and is very weak. The line width amounts to ≈20 mT which is quite low compared to the other Δ_0 resonances A, E and G. In contrast, resonance G reveals a strong increase in line width which amounts to +0.33±0.08 mT K^{-1} and is comparable to results for resonances C and D. The width is ≈45 mT at 300 K. G is involved in a double resonance with A which shows almost no temperature dependence of the line width indicating that there does not take place a change in dynamics.

The width of a static Δ_1 resonance observed in a powder spectrum amounts to ≈ 37 mT HWHM. The width of resonances B and D is obviously larger (70 mT at 300 K). The increased line width could be attributed to a disturbed geometry of the radical if the hyperfine anisotropy is enlarged. As already explained, Fleming and coworkers proposed that the cyclohexadienyl radical in NaY is bound to the sodium cation in two different configurations, endo and exo. This results in different geometries and is obvious from the muon hyperfine couplings as well. But this disturbance is expected to be too small to result in such line width effects. The large line width could also be assigned to dynamic broadening in the vicinity of the critical time scale. If so, the resonances should reveal a temperature dependence resulting in narrowing lines with increasing temperature. This is not the case, especially not for resonance D at temperatures above 300 K. Or the lines should become asymmetric due to rotational motion. In fact, resonance B was fitted with an axial line shape (D_{zz} = -12.5 MHz at 300 K). The line width for a cyclohexadienyl radical performing fast rotation about the axis perpendicular to the molecular plane amounts to -6.8 MHz. The line width of resonance B should approach this value with increasing temperature in case that the corresponding radical perfoms uniaxial rotation. As this is not the case it seems that dynamic broadening has to be excluded as well. Another possibility to explain the enlarged line width of resonances B and D would be to assume a superposition of more than one line. Remarkable is the fact that resonances C and D exhibit a considerable line broadening above 300 K but resonance B does not. The reason for this is still unclear.

4.1.3 Experimental Results and Discussion for C_6D_6/NaY

Faujasite type zeolite NaY loaded with perdeuterated benzene was investigated by Avoided Level Crossing Muon Spin Resonance (ALC-μSR) to enable the assignment of the resonances in pure NaY loaded with benzene. The exchange of the ring protons by deuterium atoms is expected to shift the proton Δ_0 lines significantly (see table 4.4), but not the muon Δ_1 and the metal Δ_0 resonances.

Table 4.4: Muon Δ_1 and proton Δ_0 resonance positions (B_{res} in T) and hyperfine coupling constants (A'_μ or A_p in MHz) of all three radicals observed in C_6H_6/NaY at 347 K (see also table 4.2). Deuteron Δ_0 resonance positions expected for C_6D_6/NaY calculated from the proton coupling constants A_p in C_6H_6/NaY.

	Δ_1			$\Delta_0(H)$		$\Delta_0(D)$
	B_{res}	A'_μ		B_{res}	A_p	B_{res}(calc.)
B	2.221	190.0	A	2.691	102.1	1.935
C	1.932	166.1	F	2.054	146.9 (296 K)	1.475
D	1.578	135.6	E	1.408	166.0	1.012

As expected the ALC-μSR spectra of C_6D_6/NaY reveal several characteristic changes compared to the non-deuterated sample (see figure 4.4). The three most prominent lines B, C and D are shifted only slightly to higher fields within a range of 20 mT which is expected for muon Δ_1 lines. The doublet at higher fields (A/G) as well as the shoulder E on the low field branch of the Δ_1 resonance D clearly disappear. Consequently, A, E and G have to be assigned to proton couplings. Therefore we exclude the assignment of Fleming *et al.* that G could be due to a sodium coupling. The Δ_1 resonances B and D both clearly reveal a shoulder that is not obvious in C_6H_6/NaY. The origin of it is still unclear. These lines could originate from sodium interactions as observed by Stolmár *et al.* in Na/ZSM-5[101] because they are shifted only slightly. An interpretation of the superimposed resonances consisting of the muon Δ_1 and the corresponding sodium Δ_0 line yields sodium hyperfine coupling constants of 50 MHz for B and 36 MHz for D. The opposite assignment gives values of more than 200 MHz and less than -120 MHz, respectively, which seems to be quite unreasonable. In calculations Webster and Macrae found a spin transfer of 12% for the sodium cation interacting with the muonated cyclohexadienyl radical which corresponds to a sodium hfc of 106 MHz.[102] No resonances were found at the Δ_0 line positions calculated for the deuterium nuclei based on the hyperfine coupling constants which were

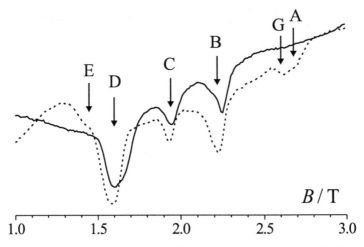

Figure 4.4: ALC-μSR spectra of C_6H_6/NaY(dashed line) and C_6D_6/NaY (solid line) at 347 K.

found for the protons. This is not surprising as the proton lines are already weak and the deuterium resonances are expected to show an even 5-fold lower intensity due to the lower gyromagnetic ratio of the deuterium compared to that of the proton (see also chapter 3.1.4). In addition the background and the width of the Δ_1 lines complicate the identification of the deuterium resonances.

4.1.4   Conclusion

We conclude that the high loading of the present sample C_6H_6/NaY, which is close to the maximum loading achievable of 5/sc and therefore bears the risk of blockage of the cages, does not influence the resonance positions in the μSR spectra. And even up to 300 K the absence of a significant temperature dependence of the Δ_1 resonance widths shows that there is no change in the dynamics. Only at temperatures higher than 300 K the widths obviously increase drastically at least for resonances C and D indicating that motional hindering increases as well. As Fleming *et al.* did not measure at higher temperatures it is not clear whether samples with loadings of 2-3/sc exhibit the same behavior in this temperature regime.

Like resonance D, line B is assumed to originate from the interaction with a sodium cation at site SII. Therefore both resonances are supposed to show comparable results. Resonance B is the only Δ_1 line which was fitted with an axial line shape. And it is not affected by the increase in line width at higher temperatures. An axial line shape with negative D_{zz} values indicates a fast rotation about the pseudo $C6$ axis of the molecule. But Macrae et $al.$ have shown that the cyclohexadienyl radical interacting with a sodium cation in NaY is rigidly fixed due to its dipole moment.[77] Looking at the ALC-μSR spectra of NaY loaded with perdeuterated benzene reveals that B could be a superposition of two lines; the second line occuring at the low field branch of the resonance leading to an apparently axial line shape. Resonance D is fitted by two different line shapes, neither of them giving an improvement of the results. This suggests that resonance D could also be a superposition of two different lines. Resonance D shows the largest line intensity of all Δ_1 resonances. Thus, it would be plausible that it consists of two contributions. In addition, the hfc of resonance D does not show a temperature dependence, whereas for resonances B and C it is obviously negative. Zeolite NaY loaded with perdeuterated benzene, C_6D_6/NaY, is expected to yield μSR spectra with Δ_1 resonances being affected only slightly but former proton Δ_0 resonances being shifted significantly. For resonance D this experiment reveals quite clearly that it consists of two different lines (see figure 4.4). In addition, the widths of resonances B and D are larger than expected for the static case. As dynamic broadening and the disturbance of the radical geometry were excluded it seems to be most likely that the larger line width has to be attributed to a superposition of lines as well.

For the three Δ_1 resonances observed in the ALC-μSR spectrum the corresponding three methylene proton Δ_0 lines were detected. Even if the hfc of resonance F is larger than predicted by Fleming et $al.$ it is most likely that this Δ_0 line corresponds to the methylene proton of the organic radical being situated at the window site of the faujasite structure. Resonance G was observed by Fleming et $al.$ as well and was suggested to be one possible candidate for the sodium Δ_0 resonance resulting, when combined with resonance B, in a hfc of around -50 MHz. This suggestion is unlikely as μSR on a sample of NaY loaded with perdeuterated benzene shows that the resonance is shifted. This indicates that it is caused by a proton not by a sodium coupling. That is, a sodium coupling as observed in C_6H_6/Na/ZSM-5 by Stolmár et $al.$ was not identified in C_6H_6/NaY up to now.

4.2   Metal Exchanged Zeolite NaY

4.2.1   Background

The interaction of paramagnetic transition metal ions with small diamagnetic molecules is usually investigated with use of EPR.[103–105] Due to their open shell nature, chemical interactions of radicals are in general expected to differ from those of comparable diamagnetic molecules. Nevertheless, for the reverse case of diamagnetic ions interacting with organic radicals only a limited number of studies is reported in literature up to now.[52, 101]

Stolmár and coworkers investigated cyclohexadienyl radicals in Na/ZSM-5 and Li/ZSM-5[101] as well as Cu/ZSM-5[52] with ALC-μSR. The observed spectra for all three systems look very similar, revealing a broad and intense resonance between 1.35 T and 1.75 T and two weak resonances, the first between 1.85 T and 1.90 T and the second between 2.10 T and 2.20 T. Comparing the spectra of the metal exchanged zeolites with the spectrum of silicalite, which is essentially free of cations but has otherwise the same framework, led to the conclusion that the resonance at intermediate field must be the Δ_1 line corresponding to the muon whereas the weak resonance at high fields was assigned to the Δ_0 line of the methylene proton. As there is no comparable resonance in the silicalite spectrum the intense resonance at lower field was interpreted as the Δ_0 line corresponding to the metal interaction. All resonance positions, the assignment and the corresponding hyperfine coupling constants are listed in table 4.6 (see the end of this chapter on page 85).

In all cases where metal cations are present the lines narrow considerably when the temperature is raised above 300 K, which indicates that the complexed species is not rigidly fixed to the zeolite lattice. The absence of narrow components in the resonances of the ionic systems shows that all benzene molecules are coordinated, whereas previous experiments at higher benzene loadings gave evidence of two component signals.[52, 106] All coupling constants are surprisingly similar for the different systems, indicating that the radical structure is affected very little by the interaction with the cation. On the basis of the isotropic hyperfine constant[85] the s-orbital spin population for the metal atom is calculated (see equation 3.7). The results are also given in table 4.6. In the case of copper the spin population is strikingly similar to the values for Ag^+ determined by Gee et al. for the same radical complexed with silver cations in polycrystalline cyclohexadiene. Gee reports the spin population for the silver, which is in the same group of the periodic table as the copper, going from 2.8 to 3.2%.[107] Stolmár and coworkers assume that the cation coordinates with the ligand via the electrons of the polarizable π-system like for the benzene and the cycohexadiene parent molecules. In addition they conclude that cation and radical find each other within a distance over which limited chemical interaction is possible. The cation content of the lithium and sodium samples amounted to nearly 2 ions per elementary cell (ec). A loading of

2.3wt.% benzene, *i.e.*, 1/ec, leads to 0.5 benzene molecules per metal ion which
is comparable to the samples discussed in the present context.

All metal exchanged zeolite samples in this work were investigated with loadings
of one benzene molecule (in the case of silver two molecules) per supercage to
avoid additional binding sites of the benzene, *i.e.*, the window sites, to simplify
analysis and thus to concentrate on the interactions of the radical with the metal
cation. Table 4.5 shows the investigated metals introduced into zeolite NaY with
their relevant isotopes, the corresponding spin, the natural abundance and the
gyromagnetic ratio used to calculate the coupling constants from the resonance
positions. In figure 4.12 at the end of this chapter (see page 86) the temperature
dependence of all resonance positions of the metal exchanged zeolites and of pure
NaY are displayed for comparison. Together with the corresponding hyperfine
coupling constants the positions are also listed in table 4.7 (see page 85).

Table 4.5: Isotopes, corresponding nuclear spin, natural abundance, gyromagnetic ratio, exchanged ion and electron configuration of the investigated transition
metal ions.

metal	isotope	spin	abund./%	γ / $T^{-1}s^{-1}$	ion	configuration
platinum	^{195}Pt	1/2	33.8	9.18×10^6	Pt^{2+}	$[Xe]4f^{14}\,5d^8$
palladium	^{105}Pd	5/2	22.2	-1.96×10^6	Pd^{2+}	$[Kr]4d^8$
silver	^{107}Ag	1/2	51.8	-1.73×10^6	Ag^+	$[Kr]4d^{10}$
	^{109}Ag	1/2	48.2	-1.99×10^6		
	Ag ^{a}			-1.86×10^6		
nickel	^{61}Ni	3/2	1.1	3.81×10^6	Ni^{2+}	$[Ar]3d^8$
zinc	^{67}Zn	5/2	4.1	2.67×10^6	Zn^{2+}	$[Ar]3d^{10}$

^{a} intermediate gyromagnetic ratio:
  $(-1.73 \times 10^6\ T^{-1}s^{-1}) \cdot 0.518 + (-1.99 \times 10^6\ T^{-1}s^{-1}) \cdot 0.482 = -1.86 \times 10^6\ T^{-1}s^{-1}$.

metal	isotope	spin	abund./%	γ / $T^{-1}s^{-1}$	ion	configuration
manganese	^{55}Mn	5/2	100	1.05×10^7	Mn^{2+}	$[Ar]3d^5$

4.2.2   AgNaY

Background

Silver exchanged zeolites are used in several catalytic and photocatalytic processes which take advantage of the presence of both, isolated silver ions and aggregated silver clusters. Examples include the photochemical dissociation of water into H_2 and O_2, the disproportionation of ethylbenzene, the oxidation of ethanol to acetaldehyde, the aromatisation of alkanes and alkenes, the selective reduction of NO by ethylene, and the photocatalytic decomposition of NO.[108,109] Lamberti et $al.$ determined the catalytically active Ag^+ sites in zeolite AgY by a combined XRPD and EXAFS study at saturation loading. They revealed that the silver cations reside at extra framework sites typical for dehydrated cations, $i.e.$, sites SI and SI' inside and next to the double-sixrings, respectively, and sites SIIa and SIIb which are both in the supercage next to the sixring window to the sodalite cage. An additional site was found to be in the middle of the sodalite cage (SI'$_m$). Lamberti and coworker conclude that a silver cation sitting here must be coordinated to residual water. The cation distribution is given to be 42% inside the supercage (sites SIIa and SIIb) and 58% outside the supercage (sites SI, SI' and SI'$_m$) not accessible to guest molecules.[108] Sites SI, SI', SIIa and SIIb for silver cations in AgY were also reported by Gellens et $al.$ with a comparable population probability of sites SIIa and SIIb.[110]

Experimental Results

Representative spectra of the silver exchanged zeolite C_6H_6/AgNaY are shown in figure 4.5. The sample was investigated in a temperature range from 223 K to 446 K with 5 mT or 10 mT step size depending on the chosen field range. At lower temperatures no significant signals were detected. The field range covered was from 0.5 T to 4.5 T. As the silver exchanged zeolite was not accessible by RFA the metal content was not determined. EA gave a benzene content of the metal loaded zeolite of 2.1wt.% benzene (1.9 molecules/sc). Presupposed the silver exchange would have been complete (maximum value: 4 Ag^+/sc) this gives a value of at least 0.5 benzene molecules per silver cation or even more in the case that the ion exchange was incomplete. This has to be expected as the analysis of the other metal ion exchanged zeolites shows.

The spectra of the silver exchanged zeolite reveal clearly two different resonances, one occuring at at 1.5899 T (446 K) being broad and intense, the other at 1.9915 T. Both resonances were fitted over a wide range of temperatures with a Lorentzian line shape becoming broader with decreasing temperature; the resonance at lower field reveals a temperature dependence with respect to the width of -0.36±0.02 mT K^{-1}, at higher field of -0.25±0.03 mT K^{-1}.

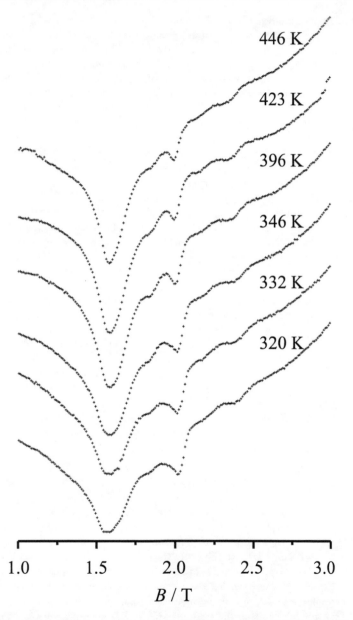

446 K

423 K

396 K

346 K

332 K

320 K

B / T

Figure 4.5: Temperature dependent ALC-μSR spectra of C_6H_6/AgNaY.

The temperature dependence of the line intensities of the two lines is different with respect to sign. Whereas the resonance at lower field increases with increasing temperature ($4.6\times10^{-5}\pm1.3^{-5}$ K^{-1}), the signal at higher field decreases ($-4.2\times10^{-5}\pm0.6^{-5}$ K^{-1}). The data are displayed in figure 4.6. Only at 446 K it was possible to fit another line of much weaker intensity at 1.8484 T. Two additional lines involved in a double resonance (2.3037 T and 2.3587 T at 446 K) are also very weak but were nevertheless detected over the whole temperature range. All resonance positions are displayed in figure 4.12 at the end of this chapter (see page 86).

According to the assignment of the resonances for Li/, Na/ and Cu/ZSM-5 the signal at 1.8484 T is interpreted as the Δ_1 line with the partner Δ_0 line at 1.9915 T. This leads to hyperfine coupling constants of A'_μ=158.3 MHz and A_p=131.2 MHz, respectively. Two isotopes of silver with spin 1/2 (^{107}Ag and ^{109}Ag) are known. Both occur in nature with approximately 50% abundance (see also table 4.5). Therefore, two different silver Δ_0 resonances are expected. Due to the only small difference in gyromagnetic ratio it might be impossible to resolve these signals (the difference in line position would amount to only 3 mT). Assuming that the signal at lower field is the Δ_0 line of the radical interaction with the silver cation shows that this obviously is the case. That is why an intermediate gyromagnetic ratio, considering the natural abundance of the two silver isotopes, is used in the following to perform a reasonable analysis of the data. The intermediate gyromagnetic ratio of -1.856×10^6 T^{-1}s^{-1} yields a silver coupling constant of 63.8 MHz. (All coupling constants are listed in table 4.7.) As the temperature dependence of the Δ_1 line is missing, it was not determined for the Δ_0 resonances. According to Stolmár and coworkers the resonances at 2.3037 T and 2.3587 T must belong to the ring protons in ortho and para position of the cyclohexadienyl radical. If this is the case the hfcs amount to 73.35 MHz and 63.17 MHz. These values are quite large for ring protons compared to values of -25.1 MHz, 7.5 MHz and -36.2 MHz for the ortho, the meta and the para ring proton in liquid benzene.[32,52]

Discussion

The ALC-μSR spectrum of C_6H_6 on C_6H_6/AgNaY is comparable to that of NaY with respect to the line widths and the line intensities; but it is obviously different with respect to the line positions. The spectrum of the silver exchanged zeolite seems to be compressed with respect to the field range (see also figure 4.12). And although the resonance at low field could be identical to resonance D observed in C_6H_6/NaY we conclude that the spectrum of AgNaY has to be assigned to a silver/radical interaction and does not originate from the sodium ions.

The muon hyperfine coupling constant (158.1 MHz at 446 K) in C_6H_6/AgNaY is reduced compared to cyclohexadienyl radicals generated from frozen benzene

(164.7 MHz at 263 K). This reduction is comparable to that found in C_6H_6/Cu/ZSM-5 (159.9 MHz) and C_6H_6/Na/ZSM-5 (160.1 MHz). Only the methylene proton coupling is slightly enlarged (131.2 MHz compared to 125.5 MHz in C_6H_6/Cu/ZSM-5). Although the silver interacting with an organic radical in the present context in embedded in a zeolite matrix the hfc of the metal cation and the deduced spin density of 3.2% is comparable to the one found by Gee *et al.*[107] who reported ESR spectra of γ-irradiated silver complexes with cyclohexadienes and 1-methyl-cyclopentene. Depending on the parent organic molecule (1,3-cyclohexadiene and 1,4-cyclohexadiene) and the parent silver salt ($AgClO_4$ and $AgBF_4$) they derived proton coupling constants of 126.1 MHz (1,4-cyclohexadiene) and 126.7 MHz (1,3-cyclohexadiene). From the silver coupling constants of 51.8 MHz ($AgClO_4$ + 1,3-cyclohexadiene) up to 61.1 MHz ($AgBF_4$ + 1,4-cyclohexadiene) they calculated spin densities on the metal cation of 2.8% and 3.2%, respectively. This confirms our interpretation that the observed ALC-μSR spectrum of C_6H_6/AgNaY originates from an interaction of the silver cation with the organic radical showing not only the muon and the methylene proton line but in addition a silver resonance.

Nevertheless, it has to be commented on the very low intensity of the muon Δ_1 resonance compared to the corresponding proton Δ_0 line. The Δ_1 resonances in ALC-μSR spectra are very sensitive indicators of anisotropic conditions. Spherical averaging of the muon anisotropy leads to a broadening of the resonance and ultimately, the loss of any meaningful intensity.[111] A rigidly fixed organic molecule in an zeolite matrix would exhibit a static line shape which is rather broad and symmetric ($\approx$37 mT HWHM). Reorientation about, *e.g.*, the axis perpendicular to the molecular plane leads to a narrower line but this time showing axial line shape. If the rotation is faster than the time scale set by $\tau_{ALC} \approx 50$ ns the hyperfine anisotropy is averaged out partially giving rise to motionally narrowed resonances. The observation of only very weak Δ_1 resonances is unambiguous evidence of nearly isotropic conditions. These could be due to fast random jumps of the cyclohexadienyl radical between tetrahedrally related sites in the cubic symmetry of the faujasite type structure. At the same time the Δ_0 lines should become narrower and lose their anisotropy, resulting in a line width comparable to that found in liquid benzene ($\approx$7 mT HWHM). In the spectrum of C_6H_6/AgNaY the width of both Δ_0 lines, the methylene proton and the silver resonance, decreases significantly. But the lines remain much broader (35 mT and 100 mT HWHM at 446 K, respectively) than in liquid benzene.

An alternative explanation for the low dip of the resonance at 1.8484 T would be that this line is misinterpreted. Assuming that the line at 1.9115 T is the Δ_1 resonance leads to a muon hfc of 170.5 MHz. The intense resonance at lowest field then yields a silver hfc of 102.8 MHz. This is much larger compared to the results of Gee *et al.* and results in an s-orbital spin density of 5.2% which is clearly out of the range determined by Gee and coworkers. Therefore, this assignment is excluded. In addition, an assignment of the low field resonance as a sodium line would yield a hfc of 105.6 MHz (ρ_s=11.8%). This value is difficult to judge as comparable data for sodium cations in zeolite NaY are missing. It is obviously much larger than the values found in NaY assuming that resonances B and D are superpositions of the muon Δ_1 and the correspoding sodium Δ_0 line (hfc: 50 MHz and 36 MHz, ρ_s: 5.6% and 4.1%, respectively; see chapter 4.1.3).

Another explanation for the large intensities of the resonances at 1.5899 T and at 1.9915 T could be the assumption that they are superpositions of the muon Δ_1 resonance and the correspoding metal Δ_0 line. This was already done in the case of NaY for resonances D and B, respectively (see chapter 4.1.3). In the case of AgNaY this leads to muon hfcs of 170.5 MHz (as already mentioned) for the high field and 136.5 MHz for the low field resonance with corresponding silver hfcs of -7.4 MHz and -5.9 MHz. An assignment of the silver resonances vice versa results in silver hfcs of 102.8 MHz (already mentioned) and -115.7 MHz. All these values seem to be unreasonable. Assuming that the resonances are superpositions with the corresponding sodium lines yields sodium hfcs of 44.9 MHz for the high field and 35.8 MHz for the low field resonance; only the latter being comparable to that found in NaY.

The discussion of the spectra of C_6H_6/AgNaY obviously shows that it is difficult to interpret the results up to now. Further information are needed to be able to explain the behavior of the cyclohexadienyl radicals in the transition metal exchanged zeolite.

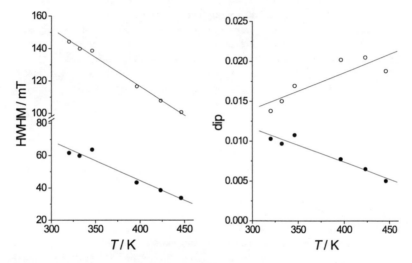

Figure 4.6: Line widths (HWHM, left) and dips (right) of the silver (open symbols) and methylene proton (solid symbols) Δ_0 resonance in $C_6H_6/AgNaY$. The solid line gives the linear fit of the data.

4.2.3   ZnNaY

Background

Zn exchanged zeolites are investigated theoretically as well as experimentally with respect to their practical use in catalysis. van Santen and coworkers, *e.g.*, undertook a theoretical study of the mechanism of ethane dehydrogenation catalysed by Zn doped zeolites. Therefore the catalyst was modelled by the ring cluster $[Al_2Si_2O_4H_8]^{2+}$ coordinated with the Zn^{2+} ion.[112] Penzien *et al.* showed that Zn^{2+} exchanged zeolite BEA was one of the most active catalysts for the cyclisation of 6-aminohex-1-yne to 2-methyl-1,2-dehydropiperidine, *i.e.*, an intramolecular addition of an amine H-NR$_2$ to a CC triple bond. The heterogeneous catalyst was more active than the corresponding homogeneous catalyst $Zn(CF_3SO_3)_2$.[113] In addition, zinc exchanged zeolites appear well suited for basic studies aimed at the determination of the concentration and characteristic adsorption behaviour of active sites in zeolites.[114,115] The literature reports investigations of the adsorption of CO in zinc exchanged Y-type zeolites, which indicate the presence of a limited number of Zn^{2+} species at or near the supercage walls with different CO adsorption capabilities.[116–118]

Experimental Results

The zinc exchanged zeolite was the only sample which could not be sealed by welding before dehydration and loading with benzene. Therefore this cell was screwed and sealed with a gasket. ALC-μSR spectra were taken at temperatures from 78 K to 448 K with 5 mT or 10 mT step size depending on the chosen field range. They are shown in figure 4.7. The total field range covered the range from 1.0 T to 3.0 T. The zinc content in the sample was determined to amount to 1.9 Zn^{2+}/sc being close to the desired maximum value. EA revealed the benzene content to be 1.6wt.% benzene, *i.e.*, 1.5 molecules/sc, leading to 0.8 benzene molecules per zinc ion which is in the optimum range of loading.

The ALC-μSR spectrum of C_6H_6/ZnNaY reveals two very intense signals at 298 K. The resonance at lower field (1.6628 T) occurs only between 223 K and 298 K. At temperatures above or below it is too broad to be fitted reasonably. In contrast to that, the resonance at higher fields (2.1731 T) can be fitted over a large temperature range although it diminishes at 373 K as well. Between these two resonances and only at 298 K a very weak line is observed which is more a shoulder on the high field branch of the low field resonance. To enable an analysis of the spectrum according to C_6H_6/AgNaY the resonance position of this line is estimated from the raw spectrum and amounts to 1.88 T. Interpreting this signal as the Δ_1 line like in C_6H_6/AgNaY in this work (see chapter 4.2.2) and in C_6H_6/Li/, C_6H_6/Na/ and C_6H_6/Cu/ZSM-5 by Stolmár due to the comparable line position yields a muon hyperfine coupling constant of 161 MHz.

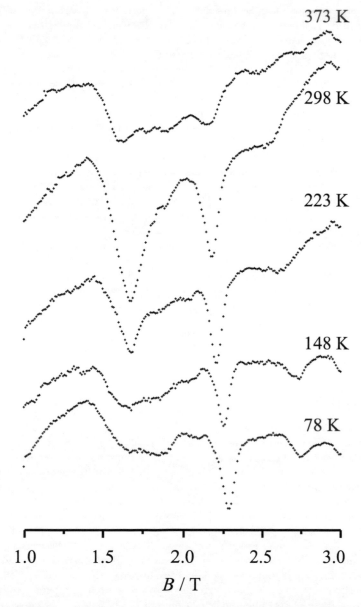

Figure 4.7: Temperature dependent ALC-μSR spectra of C_6H_6/ZnNaY.

Assuming that the high field resonance is the partner methylene proton Δ_0 line the corresponding proton hfc amounts to 106.0 MHz. Concluding that the low field resonance is due to the interaction of the organic radical with the zinc cation leads to a zinc coupling constant of 67.5 MHz. Calculating the s-orbital spin density yields a value of 3.2%. Very surprising is the fact that the high field resonance, which is the proton Δ_0 line, is the most intense line over the investigated temperature range except at 298 K. Both, in the investigations on metal exchanged ZSM-5 described in literature and the metal exchanged zeolites in the present context, the metal resonance is always the strongest signal. As for all metal exchanged zeolite samples investigated in the present context it was not possible to determine a temperature dependence of the Δ_1 resonance and therefore of the hyperfine coupling constants. The line width and the line intensity of the methylene proton Δ_0 resonance of C_6H_6/ZnNaY with respect to temperature are given in figure 4.8. The trend of both is not as clear as for C_6H_6/AgNaY. The width shows a minimum of approximately 40 mT at about 180 K, whereas the dip shows a maximum at 298 K.

Remarkable is the spectrum at 373 K. Here, the two most prominent lines are broadened so much that it is difficult to fit the single resonances in a reasonable way. In addition, the feature around 2.45 T at 298 K has to be mentioned. This certainly is an additional resonance, most likely a double resonance. As it was not possible to resolve this signal it was not investigated further. Around 3.0 T another line might be visible. Unfortunately the field range was not expanded to higher field here. Therefore, the line could not have been analysed.

Discussion

Interpreting the ALC-μSR spectrum of C_6H_6/ZnNaY in an analogous way as the spectrum of C_6H_6/AgNaY (see the previous chapter) yields a comparable muon hfc and a comparable transition metal hfc even with respect to the s-orbital spin density. Only the corresponding proton hfc is smaller than expected. Remarkable is the high intensity of the low field resonance in the spectrum of ZnNaY. In the above mentioned interpretation this signal is assigned to the interaction of the cyclohexadienyl radical with the zinc cations. But the μSR active zinc isotope ^{67}Zn amounts to only 4% natural abundance.

As already observed in the C_6H_6/AgNaY sample the muon Δ_1 resonance is surprisingly small compared to the corresponding proton Δ_0 line. This could be ascribed to motional averaging due to site hopping of the radical among the tetrahedral cation sites within a supercage as for C_6H_6/AgNaY. But in contrast, in C_6H_6/ZnNaY the width of the proton line increases with temperature (60 mT HWHM at 300 K) leading to a structure at 373 K which hardly can be resolved. Adopting the interpretation applied for pure NaY, i.e., assigning the intense high field and the low field resonance to muon Δ_1 lines superimposed by their corresponding zinc Δ_0 lines yields muon hfcs of 186.1 MHz and 142.4 MHz and zinc hfcs of 11.6 MHz and 8.88 MHz. The opposite assignment of the zinc resonances yields hfcs of 146.5 MHz and -126.1 MHz. All the zinc values seem to be unreasonable with respect to the s-orbital spin density (0.55% and 0.42%, 6.9% and 6.0%, respectively). Tracing back these lines to superpositions with sodium Δ_0 resonances gives sodium hfcs of 49.0 MHz and 37.4 MHz, being comparable but not identical to those calculated for pure NaY (see chapter 4.1.3).

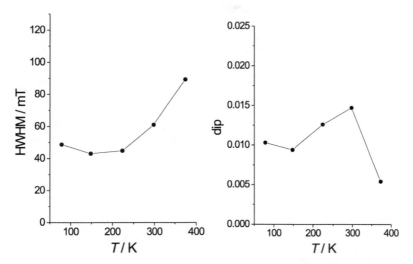

Figure 4.8: Temperature dependence of the line width (HWHM, left) and the line intensity (right) of the methylene proton Δ_0 resonance in $C_6H_6/ZnNaY$.

4.2.4   PtNaY and PdNaY

Background

Platinum and palladium supported on different types of matrices are used in various applications of heterogenous catalysis in the form of cations as well as in the form of uncharged metallic particles and clusters. Platinum and palladium supported faujasites, *e.g.*, are important catalysts in petroleum and petrochemical industries.[119] The regular channels and cages provide perfect matrices for the dispersion of metal cations and metal particles, which have been extensively characterized by IR, XPS, SAXS, EXAFS and TEM, and by catalytic tests.[120, 121] Platinum and palladium in their various oxidation states are highly active in hydrogenation, hydrocracking and dimerization reactions of small olefins.[122, 123]

Lots of investigations on both metals in zeolite matrices deal with the distribution of the ions as this is strongly dependent on the distribution of the corresponding uncharged metal particles yielded after reduction.[124–128] Before using the ions or the uncharged particles the metals have to lose their ligand shell as they are introduced mostly as amine complexes. This process is called activation. To avoid autoreduction during the activation process the calcination procedure used in the present context was performed in sufficient oxygen flow. The temperature of the activation determines the positions of the cations. To keep the platinum ions located in the supercages the calcination temperature was chosen to be lower compared to palladium.[126] In contrast to that, palladium needs higher temperatures to lose the amine ligands completely, moving step by step into smaller cavities of the zeolite system until the naked cation resides in the sodalite cage on site SI'.[125] But Ghosh *et al.* report that palladium ions located in sodalite cages have been observed migrating back towards the supercage in order to interact with adsorbates.[124]

Experimental Results for PtNaY

Representative ALC-μSR spectra recorded with the platinum exchanged zeolite C_6H_6/PtNaY in a temperature range between 121 K and 373 K are shown in figure 4.9. The step size in the field range from 0.8 T to 3.3 T was 10 mT. RFA analysis revealed a platinum content in the investigated sample of 0.75 Pt^{2+}/sc. But it revealed as well that calcination did not work properly so that the amine ligands of the parent platinum salt were still present in the sample when loading it with benzene. The benzene content was determined to be 3.8wt.% with respect to the dehydrated zeolite, that is 0.9/sc. This leads to a ratio of 1.2 benzene molecules per Pt^{2+} ion which is in the optimum range.

Looking at the spectra of C_6H_6/PtNaY is quite difficult as the background correction does not yield a horizontal base line. Therefore it is hard to distinguish between resonances and background influences. The resonance positions are estimates from the raw spectrum as fitting was not possible. Nevertheless two features are visible in almost all spectra that were recorded, one resonance around 1.9 T and another around 2.1 T. The first is broad and of large intensity at low temperatures ($\approx$120 K) and diminishes when reaching higher temperatures ($\approx$373 K). In contrast to that the line at higher field is clearly visible at higher temperatures and diminishing when cooling down the sample.

Compared with the results for metal exchanged ZSM-5 by Stolmár and coworkers the low field resonance appears in the region of the Δ_1 resonance and the high field signal in the region of the Δ_0 line; but a very broad and intense line originating from the interaction of the radical with the metal cation is not visible at all. Calculating the hyperfine coupling constants from the resonance positions leads to values of A'_μ=166 MHz and A_p=130 MHz. These values are comparable to those found by Fleming $et\ al.$ for resonance C (see chapter 4.1.2) corresponding to cyclohexadienyl radicals at the window site in NaY.

Discussion of PtNaY

RFA gave a result of 0.75 platinum ions per supercage, presupposed that the cations really all reside in the supercage. In the case that they migrate into the smaller voids upon dehydration they are no longer acessible to the benzene molecules. Instead there should occur an interaction between the cyclohexadienyl radicals and the sodium ions. But no resonance originating perhaps from a metal/radical interaction is observed. Taking into account that the calcination procedure did not work properly it is most likely that the platinum ions are still fully coordinated by amine ligands and therefore still residing in the supercages. With this ligand shell the metal cations are shielded and therefore not accessible to any other molecules. This could be the reason for the cyclohexadienyl radicals not revealing a transition metal interaction. Presupposed that the fully coordinated platinum ions block the supercages there also would be no possibility for the radical to undergo an interaction with the sodium ions. Taking into account the hyperfine coupling constants it is most likely that the cyclohexadienyl radicals reside at sites with perturbation smallest possible, $i.e.$, the window site.

Nevertheless, the line intensity in the spectra of C_6H_6/PtNaY are small compared to C_6H_6/AgNaY or C_6H_6/ZnNaY and even to C_6H_6/NaY although the benzene content is in a comparable range. As mentioned earlier platinum ions supported on different types of matrices are used as efficient catalysts. Therefore, the possibility has to be taken into account that the benzene was partially oxidized by the platinum ions. Somorjai reported on the bonding geometry of benzene in CO coadsorption structures on various metal surfaces, one of them the Pt(111).[129] He observed a significant CC bond lengthening in benzene on platinum (1.73±0.15 Å) compared to benzene in the gas phase (1.40 Å). This was even more than found for Pd(111) (1.43±0.10 Å). Thus, it seems to be very likely to us that the benzene was oxidized at least partially leading to decreased line intensities.

The reasons for the failed calcination are difficult to be found out. Samples calcined at 523 K for 4 h are thought to have completely lost the ammonia with no autoreduction and to be characterized by very disperse metallic particles in the form of Pt^{2+} which are located inside the supercages due to the low manipulation temperature.[126, 130, 131] Taken this for granted only problems with the temperature control or with oxygen supply can be responsible for the incomplete calcination.

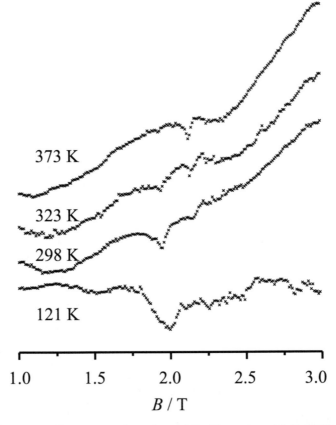

Figure 4.9: Temperature dependent ALC-μSR spectra of C_6H_6/PtNaY.

Experimental Results for PdNaY

The palladium exchanged zeolite NaY loaded with benzene was investigated in a temperature range from 123 K to 373 K with 10 mT step size. But in a field range from 1.0 T to 3.0 T no significant resonances were detected. It was not possible to determine the palladium content of this sample; but the amount of exchanged sodium cations should be comparable to that of platinum ($\approx 20\%$) as the metal ions belonging to the same group of the periodic table of the elements are comparable and the ion exchange procedure is identical. The salmon color of the sample after ion exchange suggests the presence of diamagnetic Pd^{2+}.[122]

When opening the cell after measuring μSR to determine the benzene content via EA it was obvious from the color that the sample consisted of two chemically different compounds. Approximately one third of the sample volume next to the copper tube was colored brown-gray; the rest (approximately 2/3) was ocher colored including the center of the sample cell where the muon beam is expected to focus. Elementary analysis of the ocher zeolite material revealed that the sample was loaded with only 0.6 wt.% benzene, $i.e.$, 0.1 molecules per supercage. A mixture of the the brownish-gray and the ocher material gave a value of 3.6 wt.% benzene, $i.e.$, 0.8 molecules per supercage and 86% of the theoretical maximum value expected. For neither case nitrogen was dedected. Consequently the calcination procedure worked well in this case.

From elementary analysis we conclude that the benzene did not equilibrate in the sample but got stuck directly after entering the cell so that the benzene molecules were not accessible for the muon beam. In addition the dark color of the zeolite next to the copper tube indicates that at least a part of the introduced benzene was oxidized to graphite which got stuck in the zeolite cages, coloring the sample volume only partly gray. Each event on its own inhibited the generation of cyclohexadienyl radicals which would have been observable by μSR. Probably the palladium cations on the exchanged zeolite were reduced immediately after getting in contact with benzene. At least the color of the sample during the preparation process suggests that it contained Pd^{2+} ions before loading with benzene (salmon) and Pd^{+} afterwards (ocher).[124] Bergeret and coworkers report that Pd^{2+} cations in the sodalite cages of zeolite Y can be reduced to a low valence state by benzene adsorption even though the benzene molecule cannot directly contact the cations. They conclude that the zeolite framework can act as an intermediate in the redox process between extra framework species and thus behaves like a solid electrolyte. And they point out that the overall charge transfer between the benzene molecules and the Pd^{2+} cations is in agreement with the electron donor character of the aromatic ring of benzene and with the high reduction potential of the Pd^{2+} cations.[127]

4.2.5   NiNaY

Background

Zeolites and metal oxides containing nickel are used as catalysts for ethylene dimerization. It is established that nickel is responsible for the catalytic activity, but the oxidation state of the active nickel species has been uncertain.[132] Various studies have shown that the oxidation states of catalytically active nickel species lies between 0 and +2. The coordinatively unsaturated Ni^{2+} ions on SiO_2-Al_2O_3 have been considered to be active centers.[133–135] It was also suggested that a highly dispersed Ni^0 phase in zeolites may be responsible for the catalytic activity.[136] This is in contrast to a more recent work which indicates that the activity is significantly decreased due to the reduction of Ni^{2+} ions in zeolites to 1 nm Ni particles.[137]

Experimental Results

Zeolite NaY exchanged with nickel cations was investigated in a temperature range from 173 K to 373 K. The step size in the total field range from 0.8 T to 3.7 T was 5 mT or 10 mT depending on the individual field range chosen for the single spectrum. Representative spectra are shown in figure 4.10. The metal content of the exchanged zeolite amounts to 1.4 Ni^{2+}/sc. EA yielded a benzene content of 1.2wt.% (1.1/sc).

Compared to the other spectra of metal exchanged zeolite NaY the resonances observed in the spectra of C_6H_6/NiNaY are all weak and broad. Therefore the lines were not fitted. All given resonance positions are estimates from the raw spectrum. The most intense line is observed at 1.65 T which is the field range where the metal resonances in C_6H_6/AgNaY, C_6H_6/ZnNaY, C_6H_6/Cu/ZSM-5 and C_6H_6/Na/ZSM-5 occur as well. At the high field branch of this resonance a small shoulder is visible comparable to that observed in C_6H_6/ZnNaY. The position is determined to be 1.82 T. A third resonance is detected at 2.17 T. According to literature the intermediate resonance is assigned to the muon spin flip giving a reduced muon hfc of 156 MHz. The corresponding Δ_0 proton resonance at higher fields yields a proton hfc of 90.4 MHz. Interpreting the line at low field as the nickel resonance results in a coupling constant of 58 MHz. The very low intensity of the nickel Δ_0 compared, e.g., to the silver Δ_0 resonance is not surprising as the natural abbundance of nickel with spin $\frac{1}{2}$ amounts to only 1%. And as obvious from the μSR spectra of C_6H_6/Cu/ZSM-5[52] and Na and C_6H_6/Li/ZSM-5[101] the muon Δ_1 and the corresponding proton Δ_0 resonance also weaken upon the metal interaction. The results for NiNaY are listed in table 4.7 in direct comparison to the results of the other metal exchanged zeolites.

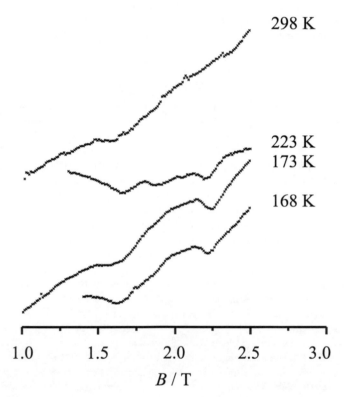

Figure 4.10: Temperature dependent ALC-μSR spectra of C_6H_6/NiNaY.

4.3   NaX

4.3.1   Background

Zeolite NaX and zeolite NaY are both built up by the faujasite framework structure. The basic difference between the two compounds is the silicon to aluminium
ratio, the latter being smaller for NaX. As the amount of Al in a zeolite structure
determines the framework charge, NaX contains more extra framework cations.
Therefore it is obvious that NaX provides at least as many cation sites as NaY
which are accessible to large adsorbate molecules. For structural details see chapter 2.4.1.

4.3.2   Experimental Results

Zeolite NaX loaded with 1.5 wt.% benzene (0.4 molecules/sc) was investigated
in a temperature range from 273 K to 348 K with 10 mT step size in a total
field range from 1.0 T to 3.2 T. At first sight the spectrum looks totally different
to that of benzene loaded NaY and shows a very interesting temperature dependence. Representative spectra are shown in figure 4.11. At low temperatures
($\leq$ 274 K) the spectra reveal four resonances: a broad line at 1.93 T, a small and
intense signal at 2.10 T and two very weak resonances at higher fields (2.92 T,
2.97 T). Rising the temperature from 274 K the resonance at low field disappears
within 1 K. The additional lines remain at their positions. Only at 346 K a small
downfield shift (in maximum by 0.05 T) is observed which is most likely due to
changes in the experimental setup. A comparable μSR investigation was done by
Fleming on zeolite NaX with a much higher benzene loading (12 wt.%).[138] At
high temperatures ($\geq$ 348 K) the spectra are identical to those found in this work
although the present loading is much smaller.

The observed spectra remind of ALC-μSR spectra on bulk benzene with resonances at 2.077 T, 2.894 T and 2.954 T.[32] Regarding the line at lowest field
as muon Δ_1 resonance yields a hyperfine coupling constant of A_μ=525.3 MHz
(A'_μ=165.3 MHz) for the muon. This result is comparable to the value found for
frozen benzene and even more with the cyclohexadienyl radical at the window
site in C_6H_6/NaY (resonance C) found by Fleming et $al.$ (compare also with
table 4.1). Resonance C also diminishes with increasing temperature. The additional hfcs amount to A_p=135.0 MHz (2.10 T) for the methylene proton and
A_p=-18.76 MHz (2.92 T) and A_p=-28.02 MHz (2.97 T) for the ortho and the para
ring protons, respectively, the first value being slightly larger, the others being
smaller than found for bulk benzene. It seems to be most likely that benzene
incorporated into zeolite NaX is situated at the window sites of the faujasite
structure.

At 275 K the muon Δ_1 resonance disappears, showing that the benzene environment changes from anisotropic ($\leq$274 K) to isotropic ($\geq$275 K) conditions. As

already mentioned, the hyperfine coupling constants of the cyclohexadienyl radical below 275 K are comparable to those found in frozen benzene. Consequently, the disappearance of the Δ_1 resonance indicates the melting point of muonated cyclohexadienyl radicals in the NaX framework. These results are comparable to those found by Auerbach *et al.* who investigated the behavior of benzene in NaX zeolite by ^{2}H NMR and molecular mechanics studies.[139] Auerbach observed a rapid rotation about the benzene 6-fold axis at temperatures below 225 K synonymous to an anisotropic environment, while the spectra recorded above this temperature reveal pseudo isotropic motion.

Up to now it is not clear why an interaction of the radical with the sodium ions like in zeolite C_6H_6/NaY is not observed although there should be more cations accessible for the organic molecules. C_6H_6/NaX was not investigated in detail as the C_6H_6/MnNaX sample did not show any resonances to be compared with.

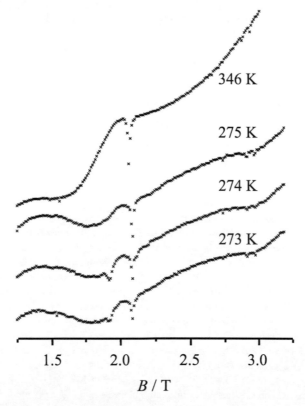

346 K

275 K

274 K

273 K

B / T

Figure 4.11: Temperature dependent ALC-μSR spectra of NaX loaded with 1.5wt.% benzene (0.4 C_6H_6/sc).

4.4 Metal Exchanged Zeolite NaX

4.4.1 MnNaX

Background

In 1999 Kim *et al.* reported the crystal structure of a benzene sorption complex of dehydrated Mn^{2+}-exchanged zeolite X with the chemical formula $Mn_{46}Si_{100}Al_{92}O_{384} \cdot 40C_6H_6$. They observed 16 mangenese ions at site SI, four Mn^{2+} at site SII and another 26 manganese ions also at site SII but chemically different. The adsorbed benzene rings were located partly next to site SII (26 molecules) interacting facially with the manganese ions and partly in the 12-ring windows (14 molecules).[96] The interaction of benzene with ions located at site SII in zeolite X was also reported by Vitale *et al.* in 1995 who investigated C$_6$D$_6$ on CaLSX. Due to the low coverage of one molecule per supercage in that study benzene rings in the 12-ring windows were not observed.[68] Qualitatively these are identical results as reported for the interaction of benzene with cations in zeolite NaY (see also chapter 2.4.1).

metal	isotope	spin	abund./%	$\gamma/T^{-1}s^{-1}$	ion	configuration
manganese	^{55}Mn	5/2	100	1.05×10^7	Mn^{2+}	[Ar]3d^5

Experimental Results

Regarding the above mentioned results manganese is the only metal being implanted into zeolite NaX instead of NaY. The basic difference between zeolite X and Y, both built up in the faujasite structure, is the silicon to aluminium ratio. As this ratio is smaller in zeolite X, the structure contains more cations to compensate the negative framework charge. A detailed description of the structure is given in chapter 2.4.1. In addition, the manganese exchanged zeolite is the only high spin system (five unpaired d-electrons, compare to table 4.5) investigated in the present context. MnNaX loaded with benzene was investigated in a temperature range from 123 K to 373 K with 10 mT step size in a field range from 0.5 T to 5.0 T. The benzene content was investigated by EA and amounts to 5.2 wt.% (1.2 molecules/sc). The metal content of the zeolite determined by RFA amounts to 2.25 Mn^{2+} ions per supercage presupposed that the metal ions really reside here. Theoretically this would be 0.5 benzene molecules per manganese ion.

Within the investigated field range C_6H_6/MnNaX did not show any resonances. Probably the manganese ions migrated into the smaller voids of the zeolite upon dehydration, being no longer accessible to the benzene molecules. Presumed the manganese ions are hidden in the smaller cages at least the benzene molecules are supposed to give resonances. There is the possibility of an interaction of the cyclohexadienyl radical with the residual sodium ions. In addition, the radicals could be situated at the window site giving resonances of a nearly unperturbed radical as observed in the present context already in C_6H_6/PtNaY (see chapter 4.2.4). Both is not the case. The concentration of benzene in the sample is sufficient. And the possibility that the benzene molecules did not equilibrate over the sample and were therefore not detected is excluded. When opening the sample cell after the μSR measurements no indications of a reaction that took place during the sample preparation procedure was observed (e.g., gray color of the zeolite due to oxidation of benzene to graphite).

The manganese ion is the only paramagnetic ion investigated in the present context. It is most likely that the interaction of the cyclohexadienyl radical with the paramagnetic manganese cation causes relaxation processes impeding the occurence of ALC-μSR resonances. As explained in chapter 2.1.1 and displayed in figure 2.1 a precondition for the appearance of ALC-μSR resonances in a 3-spin-$\frac{1}{2}$ system is that the radical electron spin remains constant. In the case that the unpaired electron of the radical undergoes interactions with the unpaired d-electrons of the manganese ions and that this relaxation process is fast enough, it is most likely that the radical electron changes its spin. Thus, the conditions for spin evaluation at the avoided level crossings is no longer given. No ALC-μSR resonances are observed.

4.5   Conclusion

4.5.1   NaY and NaX

The only structural difference between zeolite NaY and NaX is the silicon to aluminium ratio resulting in a higher framework charge and therefore in a larger number of extra framework cations in NaX. Nevertheless, the ALC-μSR spectra of the two compounds look very different. The results obtained in the present work for C_6H_6/NaY are comparable to those found by Fleming et $al.$ The Δ_1 resonances B, C and D were obtained at identical line positions. Nevertheless, resonances B and D were identified to be superpositions of two lines each. The double resonance nature of both lines appeared quite clearly in the spectra of zeolite NaY loaded with perdeuterated benzene. In addition, resonance B revealed an axial line shape in C_6H_6/NaY which was excluded by Macrae based on quantum chemical calculations. For resonance D the temperature dependence of the muon hfc and the intensity is not comparable to resonance B and C also indicating superposition. In both cases the origin of the second contribution is still ambiguous. The corresponding three Δ_0 resonances (A to B, E to D, F to C) giving reasonable methylene proton hfcs were detected as well although the coupling of F is larger than expected. The assignment of resonance G is not clear up to now. It seems to be very unlikely that G originates from a methylene proton of one of the Δ_1 resonances. In addition, the suggestion of Fleming G could be due to a sodium interaction is excluded as this resonance vanishes in a sample of zeolite NaY loaded with perdeuterated benzene. The analysis of the spectra of C_6D_6/NaY is still under way and expected to yield further information concerning the double resonance structure of B and D, and the assignment of resonances F and G. This is because the line positions of the proton Δ_0 lines, now deuteron Δ_0 lines, are affected sigificantly, whereas Δ_1 or sodium Δ_0 lines are only shifted slightly by deuteration of the organic ring.[100] Up to now no sodium/radical interaction like in C_6H_6/Na/ZSM-5 could have been identified for C_6H_6/NaY.

The spectra obtained for benzene loaded zeolite NaX remind of bulk benzene and even more of the cyclohexadienyl radical situated at the window site in zeolite NaY. This suggests that the cyclohexadienyl radicals in zeolite NaX sit at the window site and do not interact with the sodium cations. Up to now it is not clear why no metal interaction can be observed although the number of sodium cations accessible to guest molecules should be larger in NaX than in NaY. The results agree with measurements of Böhlmann et $al.$ who found chemical shifts of olefins adsorbed on zeolite NaX at loadings larger than one molecule per supercage beeing comparable to those found in the bulk material. Remarkable is the temperature dependece of the Δ_1 line which disappears within 1 K at 275 K.

4.5.2 Metal Exchanged Zeolites

The main purpose of μSR investigations on zeolites in the present context was to elucidate interactions of cyclohexadienyl radicals with extra framework transition metal cations. Therefore a number of metal exchanged zeolites was investigated, namely $C_6H_6/PtNaY$, $C_6H_6/PdNaY$, $C_6H_6/NiNaY$, $C_6H_6/AgNaY$, $C_6H_6/ZnNaY$ and $C_6H_6/MnNaX$. A direct comparison to the pure zeolites NaY and NaX is difficult as the corresponding spectra are not fully interpreted up to now (see above).

$C_6H_6/AgNaY$, $C_6H_6/ZnNaY$ and $C_6H_6/NiNaY$ show characteristic resonances. Comparison of the experimental results with data obtained for zeolite $C_6H_6/Na/$, $C_6H_6/Li/$ and $C_6H_6/Cu/ZSM$-5 (see table 4.6) lead to the conclusion that $C_6H_6/AgNaY$, $C_6H_6/ZnNaY$ and $C_6H_6/NiNaY$ show metal/radical interactions as the line widths, line intensities and line positions are in a comparable range. In addition, the line positions (except the low field resonance in $C_6H_6/AgNaY$) differ clearly from those observed in pure NaY. This is shown in figure 4.12. Therefrom we conclude that the resonances originate from a transition metal cation interaction with the cyclohexadienyl radical and not from the sodium cations.

The hyperfine coupling constants for the metal interaction of Ag, Zn and Ni are in a reasonable range as well (see table 4.7). More meaningful are perhaps the derived spin densities. In the present context they amount to values of 2.3% (Ni) up to 3.2% (Ag). The value reported for $C_6H_6/Cu/ZSM$-5 is 2.8%. Especially in the case of $C_6H_6/AgNaY$ the interpretation seems to be verified as the derived spin densities are comparable to results reported by Gee *et al.* for silver complexes with cyclohexadienes (spin densities: 2.8-3.2%). In contrast to that the values for the alkali metals are clearly higher (Na: $\approx$12%: predicted by Webster *et al.*;[102] Na: $\approx$10%, Li: $\approx$14%: calculated from ref. 101; Li: 16% from ref. 140) demonstrating that the interactions between the alkali metals and the radical are stronger than those of the transition metal cations. This is quite surprising as the alkali ions are assumed to show weaker interactions due to their larger ionic radius. It has to be taken into account that the corresponding resonances are very broad. As the spin densities for the alkali ions were calculated with the line positions obtained from the raw spectra the error is quite large.

None of the metal exchanged zeolite samples showing a transition metal/radical interaction seems to reveal a sodium/radical resonance although residual sodium ions were still present in the samples. Therefrom it is concluded that the cyclohexadienyl radicals prefer a coordination with a transition metal ion. Nevertheless, the transition metal exchanged zeolites do not show a pair of resonance sets which would indicate two different orientations of the radical as proposed for C_6H_6/NaY by Fleming *et al.* To exclude a misinterpretation of the spectra samples should have been synthesized starting from zeolite HY instead of zeolite

NaY to make sure that no sodium ions are present in the zeolite framework.

An intriguing feature in the spectra of $C_6H_6/AgNaY$ and $C_6H_6/ZnNaY$ is the fact that the proposed muon Δ_1 resonance is very weak, indicating nearly isotropic conditions. In the cubic system of the faujasite structure it might be possible that the cyclohexadienyl radicals are hopping among the four tetrahedrally arranged sites within the supercage. At the same time the corresponding proton Δ_0 resonances should decrease in width to values comparable to those observed in liquid benzene. But this is not the case for either of the two compounds. In NaY the muon Δ_1 resonances are much more intense. This indicates that there is no motional averaging due to site hopping in NaY. Consequently, the attractive force between the cylcohexadienyl radical and the metal ion must be larger in NaY than in the transition metal exchanged zeolites. This is once more surprising but was already deduced from the spin densities. But it has to be mentioned that at the same time the Δ_1 resonance is weak due to site hopping of the cyclohexadienyl radical, the corresponding metal Δ_0 resonance should decrease with respect to intensity as well. Otherwise this would mean that the whole sorption complex, $i.e.$, $[Zn^{2+}][^{\bullet}C_6H_6Mu]$ or $[Ag^+][^{\bullet}C_6H_6Mu]$, performs site hopping, not only the organic radical. The analysis of the ALC-μSR spectra of C_6D_6/NaY is still under way and supposed to deliver further information helpful for a correct interpretation.

The hyperfine coupling constants obtained from the resonance postitions in **$C_6H_6/PtNaY$** are very similar to those of unperturbed cyclohexadienyl radicals. In addition, the characteristic broad and intense signal at lower field identifying a metal/radical interaction is missing. Looking at the elementary analysis of the sample reveals that the ammonia ligands of the starting material are still present. This leads to the conclusion that in the platinum sample a metal/radical interaction was prevented by the shielding of the ammonia ligands. In addition, the $[Pt(NH_3)_4]^{2+}$ ion is too large to reside in smaller cages than the supercage. This is why the supercages are blocked for the residual sodium ions. The latter are therefore located in the smaller voids and thus not accessible for the organic radical any longer. Consequently, no sodium/radical interaction was detected either. In addition, platinum ions supported on different matrices are known as efficient catalysts. It is most likely that the benzene was partially oxidized leading to only very weak ALC-μSR resonances in the spectra of PtNaY.

$C_6H_6/PdNaY$ did not show any resonance at all. Analysing the sample revealed that on the one hand the benzene was oxidized when entering the sample cell. On the other hand it seems that the palladium was reduced at the same time. Both events prevented an interaction of Pd^{2+} ions with cyclohexadienyl radicals. Even if parts of the benzene were left in the sample there still is the possibility that the palladium ions do not reside in the supercage, but migrated into smaller cages upon losing their ligand shell. Residing in smaller voids of the zeolite the ions are no longer accessible for molecules as bulky as cyclohexadienyl radicals. $C_6H_6/MnNaX$ also did not reveal resonances in the ALC-μSR spectra. Manganese is a special case as zeolite NaX hosts more extra framework cations and therefore more cation sites. In addition, the manganese ion is the only paramagnetic metal ion being accomodated in a zeolite structure in the present context. For metal cations in NaX it is well known that they migrate to small cages upon dehydration at low metal concentrations, no longer accessible to large organic guest molecules. If this was the case there would be no explanation why $C_6H_6/MnNaX$ does not show resonances of the cyclohexadienyl radical in the X framework. We conclude that it is most likely that the interaction of the cyclohexadienyl radical with the paramagnetic manganese ion causes fast relaxation processes impeding the occurrence of μSR resonances.

Table 4.6: Literature data for metal exchanged zeolite systems loaded with benzene and investigated by μSR (resonance positions in T; hfc in MHz).

zeolite	T/K	B_{res}	A_n	B_{res}	A'_μ	B_{res}	A_p	$\rho_s/\%$
silicalite[a]	300	-	-	1.870	160.1	2.060	124.3	-
Na/ZSM-5[a]	433	1.690	86.7	1.870	160.1	2.180	102.0	10
Li/ZSM-5[a]	433	1.720	68.5	1.894	162.2	2.100	123.4	17
Cu/ZSM-5[b]	361	1.396[c]	161.0	1.867	159.9	2.049	125.5	2.8

[a]: reference [101], [b]: reference [52], [c]: ^{63}Cu.

Table 4.7: Experimental results of the transition metal exchanged zeolite systems loaded with benzene (resonance positions in T; coupling constants in MHz) obtained in the present work.

zeolite	T/K	B_{res}	A_n	B_{res}	A'_μ	B_{res}	A_p	$\rho_s/\%$
PtNaY [a]	323	-	-	1.94	166	2.13	130	-
NiNaY [a]	273	1.65	58	1.82	156	2.17	90.4	2.3
AgNaY [b]	446	1.590	63.8 [c]	1.849	158.3	1.992	131.2	3.3
ZnNaY [b]	298	1.663	67.5	1.88[a]	161	2.173	106.0	3.2

[a]: resonance positions estimated; [b]: resonance positions from fits;
[c]: obtained with the intermediate gyromagnetic ratio (see table 4.5).

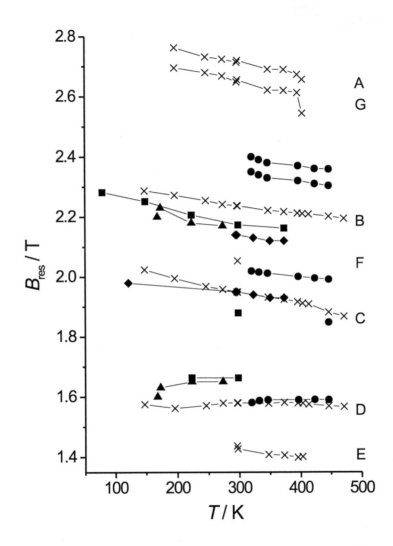

Figure 4.12: Temperature dependence of the resonance positions of all transition metal exchanged zeolite samples (C_6H_6/AgNaY: circle; C_6H_6/ZnNaY: square; C_6H_6/PtNaY: diamond; C_6H_6/NiNaY: triangle) and of C_6H_6/NaY (cross).

Chapter 5

μSR on PYRIDINIUM SALTS

Pyridinium tetrafluoroborate and pyridinium perchlorate consist of identical cations but contain different counterions. The two compounds are comparable in crystal structure and in the change of crystal symmetry with respect to temperature. But they clearly differ in the type of phase transition. In PyBF$_4$ the phase transitions are considered to be of second order,[10] whereas the perchlorate shows first order transitions.[15] As the muon is supposed to add to the unsaturated pyridinium ring these two pyridinium salts are expected to give similar resonances in μSR spectroscopy with respect to resonance position and line shape. At the same time the temperature dependence of the spectra should reveal clear differences due to the different transition types.

In principle there are four different possibilities for the muon to add to the pyridinium ring resulting in four radical species (see fig. 5.1). With respect to the nitrogen atom in the ring the muon can add to the ortho, the meta and the para carbon atom, but as well directly to the nitrogen (ipso). The different radicals will be called o-radical (ortho), m-radical (meta), p-radical (para) and i-radical (ipso) or -adduct in the following to avoid confusion when labeling the different protons within one radical. Each of these radicals is expected to give one resonance belonging to the muon (Δ_1) and one line attributed to the nitrogen atom ($\Delta_0(N)$). In addition the spectra should reveal Δ_0 resonances belonging to the methylene proton ($\Delta_0(HMu)$; the proton bound to the same atom as the muon) and to the other hydrogen atoms in ortho, meta and para position ($\Delta_0(H_o)$, $\Delta_0(H_m)$, $\Delta_0(H_p)$) and bound to the nitrogen $\Delta_0(NH)$). We expect that the strongest Δ_0 resonance will belong to the methylene protons (**HMu**), while those arising from the ortho, meta and para protons and from the proton bound to the nitrogen will be weaker, and those of nitrogen itself usually not even observed (for line intensities see also chapter 2.1.1).

In case of the o- and the m-radical there are the muon line, the methylene line and the nitrogen line and in addition five different hydrogen atoms resulting in a total number of eight resonances per radical. Due to symmetry reasons in the p- and

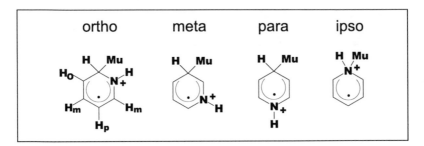

Figure 5.1: Four different radicals resulting from the addition of a muonium atom to a pyridinium ring. For the o-radical the nomenclature of the protons is shown.

in the i-adduct there are only another three different hydrogens resulting in a total number of six lines each. Consequently the ALC-μSR spectrum of PyBF$_4$ as well as of PyClO$_4$ could give theoretically in total 28 different resonances of which at least eight should be easily detectable.

5.1   Pyridinium Tetrafluoroborate

PyBF$_4$ is expected to show cation dynamics which should be detectable by ALC-μSR. In addition the dynamics is impeded continuously with decreasing temperature. This change in motion should also be visible in the ALC spectra. An interesting question is how and whether at all the second order phase transitions are revealed in the ALC measurements.

5.1.1   Experimental Results

ALC-μSR spectra on PyBF$_4$ were measured in a field range between 1.3 T and 3.8 T at temperatures between 194 K and 300 K. The step size amounted to 2.5 mT and 10 mT depending on the chosen field range and the expected resonance width. The field dependent spectrum of PyBF$_4$ recorded at room temperature is shown in figure 5.2. At first sight it reveals six large resonances between 1.8 T and 2.8 T and three smaller ones between 2.8 T and 3.8 T, all listed in table 5.1. Provided that all signals are visible and that there is no superposition of resonances in the ALC spectrum the six large resonances lead to the conclusion that there are three different radical species in the sample. But from the ALC spectrum itself it is not possible to distinguish between the corresponding Δ_1 and Δ_0(HMu) resonances. As the transverse field technique is sensitive only to Δ_1 lines it was used to verify the assignment of the resonances in the ALC spectrum.

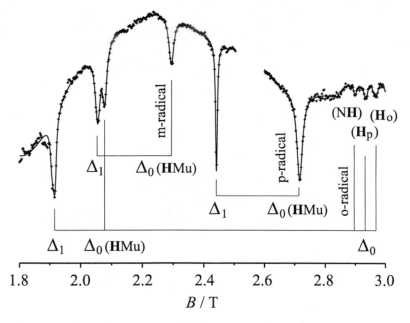

Figure 5.2: ALC-μSR spectrum of PyBF$_4$ at 300 K (crosses) and corresponding fits (solid line). The spectrum of PyClO$_4$ is identical; resonance positions differ only in the range of millitesla.

Transverse Field Measurements

Transverse field measurements on PyBF$_4$ were performed at 300 K and 0.2 T with an exponential filtering of 2.00 μs. The spectrum is shown in figure 5.3. Three pairs of different resonances with reduced coupling constants of A'_μ=163.7 MHz (A_μ=520 MHz), A'_μ=176.1 MHz (A_μ=560 MHz) and A'_μ=209.1 MHz (A_μ=665 MHz) were detected. Consequently, the ALC resonances at 1.9158 T (A'_μ=164.1 MHz, calculation see equations 3.3 and 3.4), 2.0559 T (A'_μ=176.1 MHz) and 2.4388 T (A'_μ=208.9 MHz) must be Δ_1 lines. The muon coupling constants of the Δ_1 lines are used to calculate the proton coupling constants A_p of the corresponding Δ_0(HMu) resonances from their resonance positions (see equation 3.5). For this purpose it is assumed that the sequence of Δ_0 resonance positions corresponds to the order of Δ_1 lines. The proton hyperfine coupling constants amount to 133.2 MHz, 131.5 MHz and 156.3 MHz, respectively. The ratios A'_μ/A_p of the two nuclei bound to the same carbon atom correspond to 1.23, 1.34 and 1.34, indicative of a hyperfine isotope effect that is typical for the rigid cyclohexadienyl radical.[32] As different assignment of resonances yields

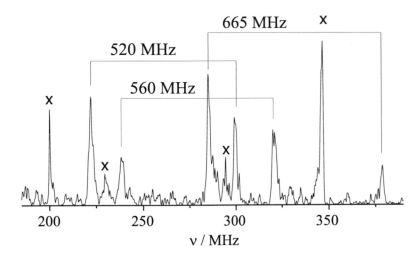

Figure 5.3: Transverse field spectrum of PyBF$_4$ at 300 K. The given frequencies are the muon coupling constants A_μ corresponding to the Δ_1 lines in the ALC spectrum. They are calculated as the sum of the single resonances of a pair. (Crosses mark artifacts.)

unlikely ratios of A'_μ/A_p it is concluded that the above mentioned identification of resonances is correct. All resonance positions, corresponding coupling constants and assignment of resonances are listed in table 5.1.

Rhodes suggested a fourth radical to be built via muonium addition to pyridinium cations, $i.e.$, the ipso-radical with muonium bound directly to the nitrogen. He derived a hyperfine coupling of $A'_\mu=156$ MHz, $i.e.$, a resonant field of 1.8177 T. As this resonance was not detected in the ALC spectra, although the other resonances are narrow and intense, this suggestion was not further investigated.[141]

Assignment of Resonances

Ronfard Haret and coworkers investigated γ-irradiation induced aza-cyclohexadienyl radicals from polyvinylpyridines (PVP) by EPR measurements.[142] The deduced coupling constants for the o-, the m- and the p-radical are given in table 5.3. In the o- and the p-radical the polyvinyl chain is bound to the pyridinium ring in the meta position with respect to the methylene group. Consequently these two meta protons are missing in the investigated compounds and therefore not detected in the EPR measurements. The m-adduct is the free aza-cyclohexadienyl radical. The coupling constant for the methylene proton in the

Table 5.1: Resonance positions, coupling constants and assignment of resonances for $PyBF_4$ and $PyClO_4$ at room temperature. (For assignment compare figures 5.1 and 5.2; for calculation of coupling constants compare chapter 3.1.4.)

	PyBF$_4$			PyClO$_4$	
Assignment μSR	B_{res}/T ALC	A/MHz [a] ALC	A/MHz [a] TF	B_{res}/T ALC	A/MHz [a] ALC
ortho Δ_1	1.9158	164.1	163.7	1.9151	164.0
meta Δ_1	2.0559	176.1	176.1	2.0543	175.9
ortho Δ_0 (**HMu**)	2.0789	133.2		2.0771	133.3
meta Δ_0 (**HMu**)	2.2928	131.5		2.2891	131.7
para Δ_1	2.4388	208.9	209.1	2.4340	208.4
para Δ_0 (**HMu**)	2.7178	156.3		2.7129	155.9
ortho Δ_0 (**NH**)	2.9049	-19.81		2.9030	-19.65
ortho Δ_0 (**H**$_p$)	2.9396	-26.24		2.9358	-25.72
ortho Δ_0 (**H**$_o$)	2.9727	-32.37		2.9707	-32.19

[a] $A = A'_\mu$ for Δ_1, $A = A_p$ for Δ_0.

Table 5.2: Resonance positions and assignment of resonances for $PyBF_4$ from μSR at room temperature and for PVP from EPR.[142] (For assignment compare figures 5.1 and 5.4.)

	PyBF$_4$		PVP
Assignment μSR	B_{res}/T ALC	Comment	B_{res}/T EPR
ortho $\Delta_0(^{14}$N$)$	1.8300	A	1.8955
meta $\Delta_0(^{14}$N$)$	2.1500	B	2.1400
para $\Delta_0(^{14}$N$)$	2.4700	C	2.3968
ortho $\Delta_0($H$_m)$		superimp. D	2.8426
meta $\Delta_0($**NH**,H$_m)$		superimp. E	2.9442
meta $\Delta_0($H$_o)$	3.1300	F	3.1622
meta $\Delta_0($H$_p)$	3.1500	G	3.1671
meta $\Delta_0($H$_o)$	3.2000	H	3.1747
para $\Delta_0($H$_m)$		not obs. I	3.5308
para $\Delta_0($H$_o)$	3.6800	K	3.6971
para $\Delta_0($**NH**/H$_o)$	3.7100	L	3.7122

o-aza-cyclohexadienyl radical derived from PVP is comparable to that found for
the radical in PyBF$_4$ with the Δ_1 resonance occuring at the lowest field (1.9158 T;
A'_μ=164.1 MHz, A_p=133.2 MHz). Calculation of the coupling constants for
the other Δ_0 lines (2.9049 T, -19.81 MHz; 2.9396 T, -26.24 MHz; 2.9727 T,
-32.37 MHz) yields also results comparable to those found by Ronfard Haret
et al. Therefore the radical with the Δ_1 resonance at lowest field in the ALC
spectrum of PyBF$_4$ is assigned as the ortho adduct. Consequently, the resonance
at 2.9049 T is the Δ_0(NH), at 2.9396 T the Δ_0(H$_p$) and at 2.9727 T the Δ_0(H$_o$)
resonance according to the assignment of Ronfard Haret *et al.* A signal for the
m-proton was not detected. This is not unexpected, since the meta couplings are
normally small, and therefore they lead to low intensity resonances. Considera-
tion of these resonances as Δ_0 lines of the m- or the p-adduct of the pyridinium
cation yields obviously differences as regards sign and magnitude so that this
assignment is excluded. As the resonances at 2.4388 T (A'_μ=208.9 MHz) and
at 2.7178 T (A_p=156.3 MHz) give the highest values for the coupling constants,
which are strikingly high, this radical species is assigned as the p-adduct as it is
found by Ronfard Haret *et al.* Thus the resonances at 2.0559 T (A'_μ=176.1 MHz)
and 2.2928 T (A_p=131.5 MHz) belong to the m-radical. This is surprising as the
proton coupling constants from μSR and EPR obviously differ although in EPR
the radical is not bound to a polyvinyl chain. No significant nitrogen resonances
were observed for any of the radicals. The missing proton lines for the m- and
the p-adduct are expected at fields higher than 3.0 T.

Simulation of ALC-μSR

With the help of a suitable simulation program, the muon coupling constant and
the proton coupling constant of the resonance of interest it is possible to simulate
the corresponding ALC spectrum with respect to line intensity and line width.
The simulation is for isotropic conditions and does therefore not show the Δ_1
resonances. The results for the three different radicals are displayed in figure 5.4.
Here, the coupling constants listed in table 5.3 marked with [‡] were used coming partly from μSR and partly from EPR investigations. The simulation of the
ALC spectrum with the help of EPR data was used to be able to investigate the
sample in more detail by μSR, *i.e.*, record spectra only in a limited field range
but therefore with smaller field steps where resonances were expected.
In addition in figure 5.4 the ALC spectrum of PyBF$_4$ at room temperature over
a broader field range is shown. The bold arrows point at resonances which were
observed in the experimental μSR spectra. Open arrows indicate field positions
where lines might be hidden behind broader resonances. The horizontal double
arrows show the field range where an expected line was not detected nor superimposed. The assignment of the additional resonances with respect to the resonant
field coming from the experiment and the simulation is listed in table 5.2. In both

Table 5.3: Comparison of coupling constants in MHz for aza-cyclohexadienyl radicals in PyBF$_4$ from ALC and in PVP from EPR.[142]

resonance		PyBF$_4$[a]			PVP[b,c]		
		ortho	meta	para	ortho[b]	meta[c]	para[b]
Δ_1	A'_μ	164.1	176.1	208.9			
Δ_0(HMu)	A_p	133.2‡	131.5‡	156.3‡	133	152	168
	A'_μ/A_p	1.25	1.36	1.35			
Δ_0(NH)		-19.81‡	-	(-28)	-19.6	10.7‡	-28.0‡
Δ_0(H$_p$)		-26.24‡	(-27)	-	-25.2	-30.6‡	
Δ_0(H$_o$)		-32.37‡	(-24)	(-22)	-33.6	-29.7‡	-25.2‡
			(-37)	(-28)		-32.0‡	-28.0‡
Δ_0(H$_m$)		-	-	-	-8.41‡	10.7‡	5.61‡
Δ_0(^{14}N)		(34)	(-13)	(6.8)	17.1‡	-9.53‡	26.1‡

[a] from ALC measurements (values in paranthesis are estimates from the raw spectrum (see fig. 5.4); other values are from fitting);
[b] from simulation of EPR spectra; [c] from spin densities.
‡ proton coupling constant used in the simulation (see fig. 5.4).

spectra the resonances already explained previously (see fig. 5.2) are marked with a star.

In the simulation the resonances for the nitrogen atoms of all three radicals are clearly visible (A: o-radical, B: m-radical, C: p-radical). They are nearly of the same intensity as the methylene lines, but very narrow. In the experimental spectrum they are hardly detectable; but their existence has been proven by the differential ALC spectrum. The N-line of the o-adduct and p-adduct are shifted by about 70 mT to lower and to higher field, respectively. The hyperfine coupling constant of 17.1 MHz for the nitrogen atom of the o-adduct derived with EPR for PVP corresponds to a line position in μSR of 1.8955 T. And indeed, in figure 5.2 it seems that at this field position a resonance is observable. But as this feature was not reproduceable it had to be attributed to beam inhomogeneities. The line for the m-adduct is found at the resonance field as simulated within the experimental error. Now the only missing resonance for the **o-adduct** is that of the meta proton. In the simulation it is observed at 2.8426 T (D) with low intensity. But in the experimental spectrum this line was not detected. In case it is shifted to lower field like the nitrogen line this signal might be superimposed by the Δ_0(Mu**H**) line of the p-adduct.

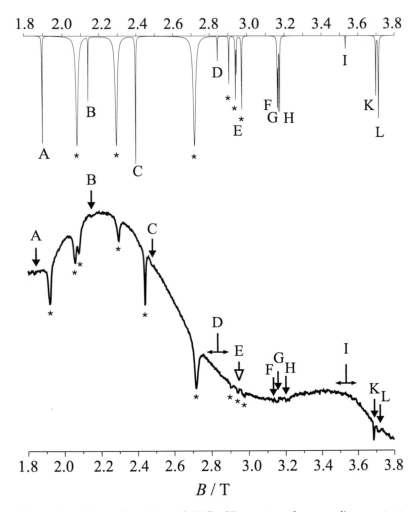

Figure 5.4: <u>Above:</u> Simulation of ALC-μSR spectrum from coupling constants determind via ALC-μSR and from EPR investigations by Ronfard Haret *et al.* (see also table 5.3).[142] <u>Below:</u> ALC-μSR spectrum of PyBF$_4$ at room temperature with broadened field range. (Bold arrow: observed resonance; open arrow: superimposed resonance (see simulation above); double arrow: expected resonance.) Resonances marked with a star were already explained in figure 5.2. The simulation is for isotropic conditions and does therefore not show the Δ_1 resonances at 1.916 T, 2.056 T and 2.439 T.

In the region between 2.9 T and 3.2 T the missing resonances of the protons of the **m-radical** are expected. But in the experimental spectrum the resonance E (for the nitrogen proton and the meta proton) is superimposed by the proton lines of the o-radical. Nevertheless, the signals for the two ortho protons and the para proton are visible although they are very weak (F, G, H). The exact assignment of the resonances is difficult, but the two lines at lower field are closer together like in the simulation. And as the EPR results were obtained from the free m-adduct it is likely that the coupling constants are identical, *i.e.*, the resonances are to be found at field positions identical to the nitrogen line. In this case F would be the resonance of the first ortho proton, G of the para proton and H of the other ortho proton of the m-radical.

Between 3.5 T and 3.8 T the additional lines of the **p-adduct** are expected, but the resonance of the meta proton (I) is again not observed. This is not suprising as it can be seen from the simulation that it is very weak. In the simulation the resonances K and L correspond to one of the ortho protons as well as to the nitrogen proton, and the other ortho proton of the p-adduct. The K line is unexpectedly intense. All additional resonances are listed in table 5.2.

In the simulation the resonances show different line widths; but even the narrow lines are very intense. Introducing relaxation processes to the simulation, *i.e.*, shortening the life time of the radicals, would have decreased the line intensity of the narrow resonances more than of the broader ones. Under these conditions the simulated spectrum would much more resemble the experimental one.

5.1.2 Analysis

The mere presence of the Δ_1 resonance shows that there are anisotropic conditions at all temperatures. All strong resonances narrow with increasing temperature, indicating increasing motional averaging. The temperature dependence of the ortho Δ_1 resonance is shown in figure 5.5. Especially remarkable is the difference between the spectra just above (239 K) and just below (238 K) the transition from the paraelectric to the ferroelectric phase at T_1=238.7 K. The Δ_1 resonance fits an axial powder line shape quite well, indicating fast uniaxial rotation. The negative sign of the axial anisotropy D_{zz}, which is related to the asymmetry of the line, shows that rotation is around the axis perpendicular to the molecular plane in all three radicals.[49] Corresponding D_{zz} values are shown in figure 5.6. Above T_1 it is around -3 MHz for the ortho and meta radicals but near -1 MHz for the para radical being nearly constant in the temperature range between 240 K and 302 K. For all three radicals the absolute value of D_{zz} above T_1 is far below its value for plain fast uniaxial rotation (-6.8 MHz for the unsubstituted cyclo-hexadienyl radical[49]) indicating extensive averaging by a tumbling, wobbling or

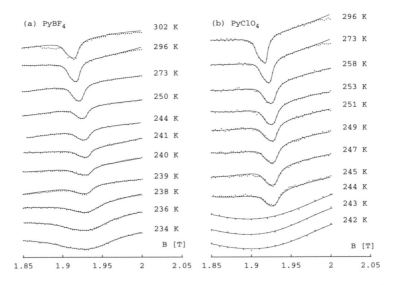

Figure 5.5: Temperature dependence of the ortho Δ_1 resonance in PyBF$_4$ (a) and in PyClO$_4$ (b) (crosses: experimental spectrum; solid line: corresponding fit).

jump reorientational motion of the rotation axis. A similar behavior was observed previously for cyclohexadienyl radicals in a high-silica ZSM-5 zeolite.[49] However, upon cooling the resonances broaden continuously until the axial powder pattern may no longer be appropriate to fit the data. Around the first phase transition temperature the D_{zz} values decrease to -8.4 MHz and -7.9 MHz for the ortho and the meta, respectively, and -6.5 MHz for the para radical. The very broad nature of the resonances at low temperature indicates that the correlation time for the rotation corresponds approximately to the critical time scale (see also chapter 2.3).

5.1.3   Discussion

The ALC-μSR spectra of PyBF$_4$ show that three different radical species - the ortho, the meta and the para radical - are formed when adding muonium. The Mu substituted aza-cyclohexadienyl radicals perform fast uniaxial rotation about the axis perpendicular to the molecular plane, confirming the results of Czarnecki *et al.*[10] Furthermore, the results give strong evidence that there is an extensive averaging by a superimposed wobbling motion of the rotational axis. The temperature dependence of D_{zz} clearly suggests a not quite abrupt suppression of the wobbling between 233 K and 244 K, paralleling a reduction of the unit cell dimensions along the rotational axis[10] (see also chapter 2.5.1) and reminiscent of the continuous nature of the phase transition found by Czarnecki.

It is remarkable that the hfc of the ortho proton in the o-aza-cyclohexadienyl radical of pyridinium tetrafluoroborate is larger than that of the para proton. This is just the opposite direction than found for unsubstituted cyclohexadienyl radicals. Obviously the nitrogen atom at one of the ortho positions significantly changes the spin population in the other ortho and in the para position of the ring.

5.2 Pyridinium Perchlorate

The $PyClO_4$ is expected to show the same features in the ALC spectra as the tetrafluoroborate concerning the cation dynamics. But clear differences should occur in the temperature dependence due to the first order transition reported in literature.[15]

5.2.1 Experimental Results

ALC-μSR measurements on $PyClO_4$ were performed in a field range between 1.3 T and 3.8 T at temperatures between 227 K and 301 K. The step size amounted to values between 2.5 mT and 20 mT depending on the field range and the expected resonance width. The spectrum of $PyClO_4$ at room temperature is identical to that of the tetrafluoroborate shown in figure 5.2. It reveals the same number of resonances, all comparable in amplitude and width. The resonance positions only differ in the range of millitesla. They are listed in table 5.1. The assignment of the resonances was done according to the analysis of the tetrafluoroborate spectrum (see chapter 5.1).

5.2.2 Discussion

The line shape of the ortho Δ_1 resonance at room temperature is equal to that in the tetrafluoroborate. Consequently the pyridinium cation performs fast uniaxial rotation about the pseudo $C6$ axis in the perchlorate as well. The temperature dependence of the ortho Δ_1 resonance is shown in figure 5.5 revealing that the line width does not change significantly down to 245 K. But below, within only 1 K, it broadens so much that a resonance is no longer detectable. Corresponding D_{zz} values are displayed in figure 5.6. They show that the line width is indeed constant above the high temperature phase transition like in the tetrafluoroborate and even 3 K below for all three radicals. But then the D_{zz} values drop due to dynamic broadening ($\tau_c \approx 50$ ns). Both, the supercooling as well as the steplike change in the D_{zz} values remind of the first order phase transition.[15] The Δ_1 line width for the meta radical below the phase transition was not determined as it overlaps with the ortho Δ_0(HMu) line (see fig. 5.2) which is broadened too much to allow a differentiation of the single contribution.

5.3   Comparison of the Compounds

As expected PyBF$_4$ and PyClO$_4$ show identical ALC-μSR spectra at room temperature because both contain the same cation. But in clear contrast the development of line shapes of the two compounds with temperature is quite different. Above the corresponding phase transition temperatures both compounds show identical D_{zz} values of the ortho Δ_1 resonance revealing that the anion does not influence the cation dynamics in the high temperature phase. The evolution of line width in the tetrafluorborate is very smooth, reminiscent of the continuous phase transition, whereas the perchlorate shows a steplike change as expected for a first order transition. In addition, in the case of the perchlorate, but not in the tetrafluoroborate, supercooling is detected, once more verifying the different phase transition types. In both cases the ortho Δ_1 resonance broadens so much that a further analysis, $e.g.$, down to the low temperature phase transition is not possible.

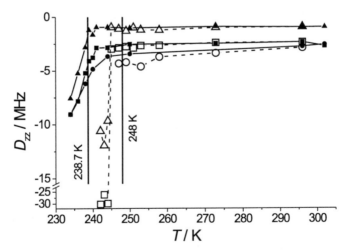

Figure 5.6: Line shape analysis of the ortho Δ_1 resonance in PyBF$_4$ (solid symbols) and PyClO$_4$ (open symbols) for all three radicals (ortho: square; meta: circle; para: triangle).

Chapter 6

NMR on PYRIDINIUM TETRAFLUOROBORATE

In the following solid state ^{2}H NMR investigations on perdeuterated pyridinium tetrafluoroborate (PyBF$_4$-d$_5$) in the temperature range between 120 K and 290 K are presented.[143] Special interest lies in the behavior of the cation molecule in the vicinity of the two solid-solid phase transitions and in the low temperature phase as the structural characterization of PyBF$_4$ in the latter is still fragmentary. The quantitative analysis of the line shape and relaxation experiments provides a detailed insight into the cation dynamics as a function of temperature and therefore of phase. Moreover, the analysis leads to kinetic and thermodynamic information. As these data only relate to the cation, comparison with experimental results for the macroscopic properties from differential thermal analysis[11] or measurements of permittivity, pyroelectric effect (in the following: pyroeffect) and spontaneous polarization[10] leads to detailed information about the individual ionic contribution. Thus it is possible to determine the intermolecular interactions which lead to ferroelectricity. Moreover the fraction of ferroelectricity caused by the classical mechanism and the contribution caused by the onset of ordering of the ions can be distinguished.

6.1 ^{2}H NMR Spectra

6.1.1   Experimental Results

All experimental ^{2}H NMR spectra of the powder sample measured in the temperature range between 120 K and 290 K covering all three phases are shown in figure 6.1. The overall shape of the temperature dependent spectra shows a typical solid state ^{2}H NMR powder pattern, so called Pake type pattern. It is obvious that both phase transitions have a strong impact on the corresponding line shapes. At 204 K and at 205 K as well as at 238 K superposition of two different spectra is observed. In general, the distinct features (*i.e.*, sharp singularities) of these ^{2}H NMR spectra imply the presence of motions that are in the fast exchange limit on the NMR time scale. This was further verified by additional T_2 experiments.

In the high temperature phase ($T > 238.7$ K) axially symmetric line shapes are detected. The observed splitting between the perpendicular singularities of $\Delta\nu_1 = \Delta\nu_2 = \Delta\nu_\perp = 65.3$ kHz is independent of temperature and considerably reduced as compared to the "rigid limit" value. These observations clearly indicate the presence of a highly symmetric and efficient motional process. Cooling the sample below the phase transition at $T = 238.7$ K results in typical non-axially symmetric ^{2}H NMR line shapes. In the intermediate phase the overall ^{2}H NMR pattern is found to vary continuously with the actual sample temperature down to a value of $\Delta\nu_1 = 1.98$ kHz at $T = 220$ K. Just below this the minimum splitting is observed, followed by a slight increase up to the second phase transition at $T = 204$ K. Again, the experimental line shapes can be understood on the basis of a fast motional process which now must be of lower symmetry. The low temperature phase transition at $T = 204$ K is reflected in the ^{2}H NMR line shapes by a sudden change of the spectral splitting. Upon further cooling to 120 K an almost "rigid limit" powder pattern is observed. Now, the experimental splitting between the perpendicular singularities is given by 125 kHz, a value that is expected for an almost completely immobile pyridinium cation. The experimental splittings between the spectral singularities, as derived from the variable temperature dependent ^{2}H NMR line shapes, are summarized in table 6.1.

6.1.2   Simulations

3-site Jump Model

The simulated ^{2}H NMR spectra are also given in figure 6.1. They are obtained by modeling the cation reorientational motion on the basis of a 3-site jump model. Comparison of the experimental and the simulated line shapes shows that the chosen simulation model is appropriate to describe the experimental spectra very well. In the high temperature phase equally populated sites ($p_1 = p_2 = 0.33$, $i.e.$, $\Delta G° = 0$) are found, whereas below 238.7 K p_1 increases at cost of p_2 upon cooling. The evaluation of the population with temperature displayed in figure 6.2 confirms quite clearly the dramatic changes near the phase transitions. The p_1 values are given in table 6.2.

6-site Jump Model

As expected, on the basis of the 6-site jump model the same good reproduction of the experimental spectra was achieved as found for the 3-site jump case. The values of the population of the lowest energy orientation p_1 are also given in table 6.2 and displayed in figure 6.2. Obviously the values for p_1 are identical

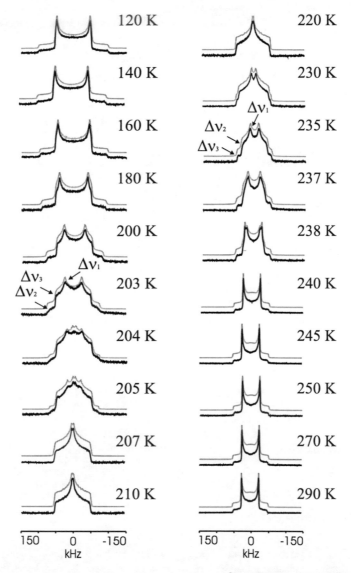

Figure 6.1: Temperature dependent experimental ^{2}H NMR spectra of PyBF$_4$-d$_5$ (black) and simulated spectra (gray) obtained from a 3-site jump model. The simulations are offset for ease of comparison and absolutely identical when using a 6-site jump model. The population values p$_1$ for the 3-site and the 6-site jump model are given in table 6.2 and displayed in figure 6.2.

Table 6.1: Experimental splittings of all ^{2}H NMR spectra of PyBF$_4$-d$_5$ (compare also fig. 6.1).

T/K	$\Delta\nu_1$/kHz	$\Delta\nu_2$/kHz	T/K	$\Delta\nu_1$/kHz	$\Delta\nu_2$/kHz
120	125	265	220	1.98	127
140	123	261	230	15.8	111
160	121	254	235	31.7	91.1
180	109	242	237	47.5	83.2
200	79.2	230	238*	51.5	79.2
203	63.4	194	238**	65.3	79.2
204*	55.5	190	240	65.3	65.3
205*	51.5	186	245	65.3	65.3
204**	7.92	190	250	65.3	65.3
205**	7.92	186	270	65.3	65.3
207	5.94	143	290	65.3	65.3
210	3.96	139			

$\Delta\nu_3 = 131$ kHz = const.
* 1st spectral component, ** 2nd spectral component.

for both models in the low and almost identical in the intermediate temperature phase including the characteristic changes at the phase transition temperatures. In the high temperature phase equally populated sites are found ($p_1 = 0.167$). This leads to a more steplike change at the high temperature transition compared to a nearly continuous change for the 3-site jump model.

2-site Jump Model

Simulation of the temperature dependent spectra with a 2-site jump model yields qualitatively equal results. The corresponding value to the population in the 3-site and 6-site jump model is the jump angle which covers a range from 90° in the high temperature phase down to 20° at 120 K. Qualitatively, it shows the same trend with temperature like the population obtained from the the other models

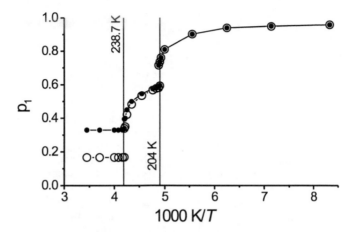

Figure 6.2: Population p_1 of the lowest energy orientation obtained from the simulation with a 3-site jump model (solid circles) and with a 6-site jump model (open circles). Phase transitions are indicated with vertical solid lines (literature values see ref. 10).

6.1.3   Discussion ^{2}H NMR Spectra

The experimental ^{2}H NMR spectra of PyBF$_4$-d$_5$ are found to be in the fast motional limit, which is true for the whole temperature covered in the present study. In particular this also holds for the low temperature phase where correlation times larger than 10^{-8} s are observed. Here, the population p_1 is close to unity; the quadrupolar interaction therefore is only slightly modulated by the underlying motional process of the cations, as shown also by independent line shape simulations. The assumption of the fast motional limit is therefore justified. Except at the phase transition temperatures no superimposed spectra are found. Therefrom it is concluded that all deuterium atoms in the pyridinium molecule must be equivalent concerning the cation motion. Therefore, a rotation around a $C2$ axis in the molecular plane was excluded. Hence, the axial line shape in the high temperature phase is understood assuming a fast rotation about the pseudo $C6$ axis, the molecular plane normal. It coincides with the [111] crystal axis which is the unit cell diagonal (see fig. 2.4). Besides, the spectra of N-deuterated PyBF$_4$ are identical, showing once more that the rotational motion takes place exclusively about the $C6$ axis. The non-axial spectra in the intermediate temperature regime allude to a non-degenerate $C3$ or $C6$ jump rotation in the fast exchange limit. At 120 K, a nearly "rigid limit" powder pattern is observed, due to a nearly static molecule perhaps only performing a small oscillation. These changes in line shape show very clearly that the fast rotational motion in the high temperature

Table 6.2: Population p$_1$ of the lowest energy orientation obtained with 3-site and 6-site jump model.

T/K	3-site	6-site	T/K	3-site	6-site
120	0.958	0.958	220	0.545	0.534
140	0.950	0.950	230	0.500	0.483
160	0.940	0.940	235	0.450	0.422
180	0.902	0.902	237	0.400	0.351
200	0.810	0.810	238*	0.395	0.340
203	0.760	0.760	238**	0.340	0.168
204*	0.735	0.735	240	0.333	0.167
205*	0.715	0.715	245	0.333	0.167
204**	0.595	0.593	250	0.333	0.167
205**	0.585	0.581	270	0.333	0.167
207	0.580	0.577	290	0.333	0.167
210	0.575	0.567			

* 1st spectral component, ** 2nd spectral component.

phase becomes more and more hindered upon cooling until it is nearly stopped. The most distinct changes in motion are found to take place around the phase transitions. The spectra at 204 K, 205 K and 238 K (see fig. 6.1) are the only examples with two pairs of main singularities each (see also table 6.1) showing that at these points identical motion with different rate constants takes place simultaneously. This may indicate a temperature gradient of one degree in the sample. Here, the symmetry of the system changes, and at the high temperature transition this leads to a slight compression of the [111] axis which limits the space between the individual cation layers. This effect could be responsible for the hindering of the rotational motion. Comparison of the splittings and the line shapes with the results for perdeuterated benzene[144] supports this interpretation and leads to the same conclusions concerning the temperature dependence of the rotational motion of the pyridinium cation in PyBF$_4$ as suggested previously by Czarnecki et al.[10]

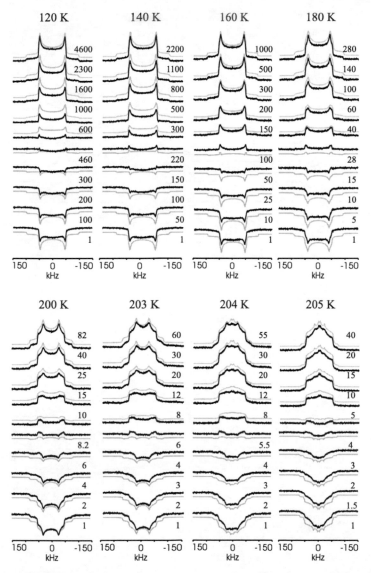

Figure 6.3: Partially relaxed experimental ^{2}H NMR spectra of PyBF$_4$-d$_5$ (black) and simulated line shapes (gray) obtained from a 3-site jump model in the **low temperature regime**. Simulations are slightly offset to facilitate comparison and absolutely identical when using a 6-site jump model. Numbers on the right give the delay times in milliseconds.

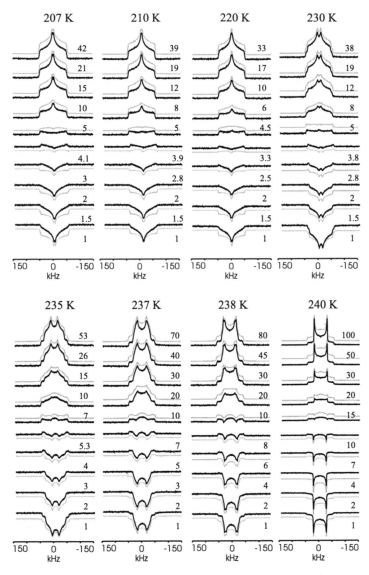

Figure 6.4: Partially relaxed experimental ^{2}H NMR spectra of PyBF$_4$-d$_5$ (black) and simulated line shapes (gray) obtained from a 3-site jump model in the **intermediate temperature regime**. Simulations are slightly offset to facilitate comparison and absolutely identical when using a 6-site jump model. Numbers on the right give the delay times in milliseconds.

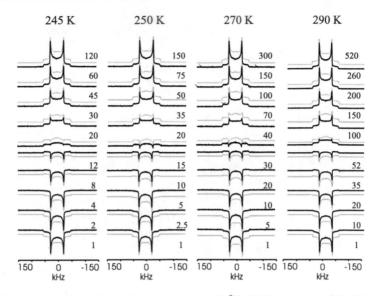

Figure 6.5: Partially relaxed experimental ^{2}H NMR spectra of PyBF$_4$-d$_5$ (black) and simulated line shapes (gray) obtained from a 3-site jump model in the **high temperature regime**. Simulations are slightly offset to facilitate comparison and absolutely identical when using a 6-site jump model. Numbers on the right give the delay times in milliseconds.

6.2 Relaxation Experiments

6.2.1 Experimental Results

All experimental partially relaxed spectra are displayed in figures 6.3, 6.4 and 6.5. T_{1Z} values determined from the experimental FID are plotted in figure 6.6. They show a very characteristic curve covering the range from around 5 ms to 800 ms. The T_{1Z} minimum is found at $T=220$ K, indicating that the motion of the cation occurs around the Lamor frequency of 46.07 MHz. The two phase transitions could also be determined from the T_{1Z} curve as the slope changes at the transition temperatures. The T_{1Z} curve is qualitatively comparable to those obtained from ^{1}H NMR measurements on pyridinium perchlorate by Czarnecki et al.[15] and on pyridinium fluorosulfonate by Pajak et al.,[25] although the latter shows three and not only two phase transitions. At very low temperatures the slope of the curve in the present context is not constant, which has been verified not to be due to saturation effects.

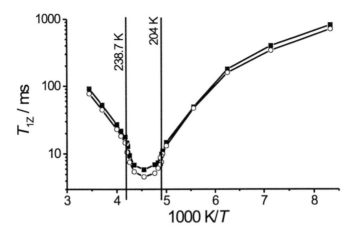

Figure 6.6: Experimental (solid squares) and simulated (open circles; 3-site jump model) spin-lattice relaxation time T_{1Z}. Phase transitions are indicated with vertical solid lines. The simulated values are absolutely identical when using a 6-site jump model.

6.2.2   Simulations

3-site Jump Model

The simulated partially relaxed spectra obtained from the 3-site jump model are also displayed in figures 6.3, 6.4 and 6.5. They show very good correspondence to the experimental ones. Theoretical T_{1Z} values result from the best fit. Figure 6.6 shows that the overall charactristics of the T_{1Z} curve is also simulated very well. Simulation of the partially relaxed line shapes in addition yields correlation times τ_c covering the range from 2×10^{-7} s to 3×10^{-11} s, and corresponding to rate constants

$$k = \frac{1}{\tau_c} \tag{6.1}$$

which are displayed in figure 6.7.

The rate constant k gives the number of all individual jumps per unit time, which is the sum of the individual rate constants k_{12} and k_{21} representing a jump out of and into the highest populated orientation, respectively.[145, 146]

$$k = 2k_{12} + k_{21} \tag{6.2}$$

$$k_{12} = 1/\tau_{12} = p_2 \; k = p_2/\tau_c \tag{6.3}$$
$$k_{21} = 1/\tau_{21} = p_1 \; k = p_1/\tau_c$$

Arrhenius behavior of the rate constants is found in all three phases. In the intermediate phase a point is found outside the regression line, which stems from the discrepancy between the experimental and theoretical T_{1Z} data, as mentioned earlier. Likewise, in the low temperature phase the rate constants at 120 K and 140 K do not follow Arrhenius behavior which was already observed in the underlying experimental T_{1Z} data. This latter observation might be traced back to the presence of an additional mechanism that contributes to ^{2}H spin relaxation at low temperatures, as for example by the motion of the tetrafluoroborate anion. According to the Eyring formulation

$$E_A^{\neq} = -R \left(\frac{\delta \; ln(1/\tau_c)}{\delta(1/T)} \right)_p \tag{6.4}$$

the activation energies $E_A^{\neq}$ calculated from the slopes in the plot of the combined rate constant, $k = 2k_{12} + k_{21}$, are 19.3 kJ mol^{-1}, 46.4 kJ mol^{-1} and 10.0 kJ mol^{-1} in the high, the intermediate and the low temperature phase, respectively.
Very similar values of 17.4 kJ mol^{-1} for PyBF$_4$ and of 17.6 kJ mol^{-1} for pyridinium iodide in the high temperature phase were found by Wasicki et al.[147] and by Ripmeester,[8] respectively, both being only slightly lower than found in the present context. Likewise, the value for the high temperature phase is comparable to that of a $C6$ rotation in pure solid benzene (16.5 kJ mol^{-1}) reported by Ok et al.[144] Interestingly, $E_A^{\neq}$ shows a maximum in the intermediate temperature regime. The corresponding frequency factors

$$A = \left(\frac{1}{\tau_c} \right)_{T \to \infty} \tag{6.5}$$

amount to 8.9$\times$10^{13} s^{-1}, 6.1$\times$10^{19} s^{-1} and 3.7$\times$10^{10} s^{-1} with the intermediate value being once more unexpectedly high. Obviously the activation energy and the frequency factor are strongly correlated. Comparable high frequency factors were reported by Gullion et al. for diffusion processes in benzene.[148]

Transition State: Eyring

The enthalpy of the transition state is calculated from the activation energies $E_A^{\neq}$ according the Eyring formulation:

$$\Delta H^{\neq} = E_A^{\neq} - RT \tag{6.6}$$

From the axis intercept A in the Arrhenius plot of the correlation time the transition state entropy is determined:

$$
\begin{aligned}
\Delta S^{\neq} &= \frac{R}{\log e}\left[\log\left(\frac{1}{\tau_c^0}\right) - \log\left(\frac{k_b T e}{h}\right)\right] \\
&= R\left[\ln\left(\frac{1}{\tau_c^0}\right) - \ln\left(\frac{k_b T e}{h}\right)\right] \\
&= R\left[\ln A - \ln\left(\frac{k_b T e}{h}\right)\right]
\end{aligned}
\tag{6.7}
$$

The Gibbs free energy of the transition state amounts to:

$$\Delta G^{\neq} = \Delta H^{\neq} + T\,\Delta S^{\neq} \tag{6.8}$$

The activiation energies $E_A^{\neq}$ give only mean values for a jump out of and back into the p$_1$ orientation. To be able to differentiate between the single jump processes, the activation energy, enthalpy, entropy and Gibbs free energy for a single process are calculated. Bovey et al.[145] performed this analysis for two different conformers tantamount to a 2-site jump process. Adjusting this analysis to a 3-site jump process yields for the single enthalpy:

$$\Delta H_{12}^{\neq} = \Delta H^{\neq} + p_1\,\Delta H^{\circ} \qquad \text{and} \qquad \Delta H_{21}^{\neq} = \Delta H^{\neq} - 2p_2\,\Delta H^{\circ} \tag{6.9}$$

and therefrom for the single activation energy:

$$E_{12}^{\neq} = \Delta H_{12}^{\neq} + RT \qquad \text{and} \qquad E_{21}^{\neq} = \Delta H_{21}^{\neq} + RT \tag{6.10}$$

$$
\begin{aligned}
E_{12}^{\neq} - E_{21}^{\neq} &= \Delta H_{12}^{\neq} - \Delta H_{21}^{\neq} \\
&= (p_1 + 2p_2)\,\Delta H^{\circ} \\
&= \Delta H^{\circ}
\end{aligned}
\tag{6.11}
$$

The index $_{12}$ indicates the jump process out of the p$_1$ into the p$_2$ or p$_2'$ orientation, whereas index $_{21}$ indicates the jump back into the p$_1$ orientation. The Gibbs free energy is calculated as follows:

$$\Delta G_{12}^{\neq} = \Delta G^{\neq} + RT \ ln\left[1 + exp\left(\frac{\Delta G^{\circ}}{RT}\right)\right] = \Delta G^{\neq} + RT \ ln\left(\frac{p_1}{p_2} + 1\right) \quad (6.12)$$

$$\Delta G_{21}^{\neq} = \Delta G^{\neq} + RT \ ln\left[1 + exp\left(-\frac{\Delta G^{\circ}}{RT}\right)\right] = \Delta G^{\neq} + RT \ ln\left(\frac{p_2}{p_1} + 1\right)$$

$$\begin{aligned}
\Delta G_{12}^{\neq} - \Delta G_{21}^{\neq} &= RT \ ln\left(\frac{p_1}{p_2} + 1\right) - RT \ ln\left(\frac{p_2}{p_1} + 1\right) \quad (6.13) \\
&= -RT \ ln\left(\frac{p_2}{p_1}\right) = -RT \ lnK = \Delta G^{\circ}
\end{aligned}$$

For the calculation of ΔH° see chapter 6.5, for ΔG° see also chapter 3.2.2.

As already mentioned the activation energies $E_A^{\neq}$ derived from the 3-site jump simulation give only mean values for a jump out of and back into the p_1 orientation. To be able to differentiate the two jump processes it is necessary to apply the Eyring theory. Therefore, the latter was derived from a 2-site jump case and assumed to be applicable to a 3-site jump process. It will be shown in the following that the 6-site jump process is probably even more appropriate model to describe the investigated system. However, extending the Eyring theory to the 6-site jump case would have been even much more complex. But this theory was not used for the data analysis worked out in the present context. Therefore, it is only described theoretically for the 3-site jump case and not applied to the experimental NMR data. If the investigations on pyridinium salts are followed up, *e.g.*, performing NMR investigations on PyClO$_4$, it is certainly worth to apply the Eyring theory on the experimental data and perhaps even to extend it to the 6-site jump case.

6-site Jump Model

Simulating the partially relaxed spectra with a 6-site jump model yields identical line shapes and an identical T_{1Z} curve as obtained with the 3-site jump model (see also figures 6.3, 6.4, 6.5 and 6.6). Only the values for the rate constants are shifted upward but show the same trend including one point out of the regression line and a non-linear behavior at low temperatures. The values are shown in figure 6.7 and cover a range from 1×10^7 s to 1×10^{11} s. As for the 3-site jump model the individual rate constants are calculated as:

$$k_{ij} = \frac{p_{ij}}{\tau_c} = p_{ij}\ k \qquad\qquad (6.14)$$

The activation energy derived from the correlation times assuming a 6-fold jump process amount to 19.1 kJ mol^{-1}, 54.0 kJ mol^{-1} and 11.2 kJ mol^{-1} in the high, the intermediate and the low temperature phase, respectively (see equation 6.4). The frequency factors are 3.3×10^{14} s^{-1}, 1.2×10^{22} s^{-1} and 1.8×10^{11} s^{-1} (see equation 6.5).

6.2.3 Discussion of Relaxation Experiments

In general, the simulated line shapes provide a very good reproduction of their experimental counterparts. The theoretical partially relaxed ^{2}H NMR spectra obtained on the basis of the aforementioned 3-site jump model are indentical to those obtained from the 6-site jump model. The same holds for the theoretical T_{1Z} values. The temperature dependence of the rate constants is also almost identical resulting in comparable values for the activation energy. Only the absolute values for the 6-site jump model are shifted upward giving higher frequency factors on the basis of this simulation model. The unreasonably high parameters of activation energy and hence the frequency factor in the intermediate phase for both models point to the presence of a composite process. If, for an alternative analysis, the high temperature frequency factor is assumed to remain valid also for the intermediate phase the experimental correlation times can be matched also with a stepwise increase of the activation energy from 19 to 24 kJ mol^{-1} over the intermediate phase. It reflects the way $\Delta G°$ (and thus $V_1°$, see also chapter 3.2.2) is built up when the high temperature symmetry is broken and the pyridinium ions experience a preferential orientation in the electric field that is provided by the macroscopic polarization.

The analysis of the variable temperature ^{2}H NMR data on the basis of both models reveals characteristic changes of the pyridinium properties, *i.e.*, relaxation times, correlation times, population of jump sites, in the vicinity of the two solid-solid phase transitions. A simple 2-site jump model, accounting for oscillatory motions, can be ruled out, since it failed for the description of the inversion recovery measurements.

6.3 Ferroelectric Polarization

The polarization is defined as the average dipole moment per unit volume. The orientational contribution can be calculated from the temperature dependent population of the cation orientations, providing that the molecular electric dipole moment of the cation is known.

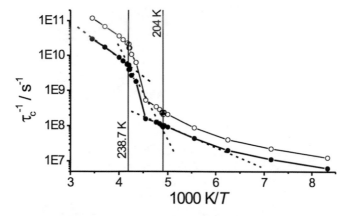

Figure 6.7: Arrhenius plot of the inverse correlation time τ_c obtained from simulation with a 3-site jump model (solid circles) and with a 6-site jump model (open circles). Phase transitions are indicated with vertical solid lines. Dashed lines represent the slopes of the curves for the 3-site jump model serving to estimate activation energies $E_A^{\neq}$.

6.3.1 Ab Initio Calculation of the Dipole Moment

To determine the electric dipole moment quantum chemical calculations using Gaussian98 based on B3LYP functionals with various basis sets were undertaken. They revealed that the dipole moment is nearly basis set independent. It amounts to 1.876±0.006 D for a simpler (6-31G*) and 1.811±0.004 D for a more sophisticated (6-311++G(2df,2pd)) basis set. The use of augmented basis sets did not introduce any further change. Calculations for the pyridine molecule showed that the results for the simpler basis sets are very close to the literature value. Therefore the first value (1.876 D) was chosen for further analysis. The dipole moment is situated in the molecular plane of the pyridinium cation and points at the nitrogen atom. The center of the cation charge is in general not expected to coincide with the cation rotation axis. When the orientational order increases, this would lead to an additional contribution to polarization of the displacement type of positively and negatively charged sublattices, without actually displacing the ion center of masses. To check for this effect, the dipole moment was also calculated for the pyridinium cation located in an anion matrix. This matrix consisted of eight point charges (-0.125 |e| each), arranged around the cation according to the low temperature unit cell structure. Rotation of the cation shows that the effective dipole moment is indeed slightly orientation dependent, but the variations are too small (<2%) to show a significant effect on the polarization.

basis set	μ_{PyH^+}/D	error/D [a]
6-31G*	1.876	±0.006
6-31G**	1.872	±0.006
6-311G*	1.896	±0.006
6-311++G(2df,2pd)	1.811	±0.004
aug-cc-pVDZ	1.797	±0.004
aug-cc-pVTZ	1.800	±0.004
aug-cc-pvQZ	1.799	±0.004

[a] error obtained from Gaussian98

6.3.2   Polarization

The population averaged dipole moment per cation, as determined from the NMR data, times the number of the cations per unit volume, then gives the orientational contribution to the absolute polarization.

3-site Jump Model

In case of the 3-site jump model the absolute polarization amounts to:

$$pol_{NMR} = \mu_{PyH^+} (p_1 - 2p_2 \cdot cos(60°)) \qquad (6.15)$$
$$pol_{NMR} = \mu_{PyH^+} (1.5p_1 - 0.5)$$

It is shown in figure 6.8 along with the macroscopic polarization reported from pyroeffect measurements by Czarnecki *et al.* on a PyBF$_4$ single domain single crystal.[10] Due to identical population of the three orientations there is no polarization in the high temperature phase. In the intermediate phase the values rise to almost 1.4 μC cm^{-2} at 204 K for the data from both the results of NMR and of pyroeffect measurements, and the two curves agree almost within error. However, in the low temperature regime the polarization from the pyroeffect shows only a slight further increase (see inset of fig. 6.8), whereas the values derived from NMR data rise to 3.6 μC cm^{-2}, which is more than double the value at 204 K (see also listing of results in table 6.3). The NMR data represent only

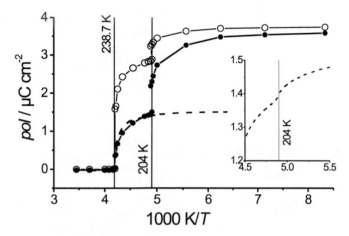

Figure 6.8: Polarization calculated from population for the 3-site jump model (solid circles), for the 6-site jump model (open circles) and spontaneous polarization obtained from pyroeffect measurements (dashed line) by Czarnecki *et al.*;[10] the inset shows an expansion of the macroscopic polarization around the low temperature phase transition.

the cation rotational order. The good agreement between the two curves implies that in the intermediate temperature regime the ordering of the cations is within the validity of the 3-site jump model the only mechanism that contributes to the macroscopic polarization. On the same basis, the discrepancy between the two curves in the low temperature regime reveals that an additional mechanism with an opposite effect contributes to the polarization. The polarizability of the pyridinium cations is too small to account for this, and moreover it should operate also in the intermediate temperature phase. It seems that only the classical ferroelectric mechanism, a translational displacement of cation versus anion sublattice, is left as a reasonable origin of the discrepancy.

6-site Jump Model

For the 6-site jump model the absolute polarization is calculated as follows:

$$pol_{\text{NMR}} = \mu_{\text{PyH}^+} (p_1 + 2p_2 \cdot cos(60°) - 2p_3 \cdot cos(60°) - p_4) \qquad (6.16)$$
$$pol_{\text{NMR}} = \mu_{\text{PyH}^+} (p_1 + p_2 - p_3 - p_4)$$

It is also shown in figure 6.8. As for the macroscopic data and for those derived from the 3-site jump model there is no polarization found in the high temperature phase. But in the intermediate temperature regime the polarization already rises to a value of 2.9 μC cm^{-2}. This is 60% of the total value of 3.8 μC cm^{-2} which is attained at 120 K. Although the value reached in the intermediate phase on the basis of the 6-site jump model is much higher than for the macroscopic data the shape of the curve is identical.

6.3.3   Sublattice Displacement

3-site Jump Model

Assuming a 3-site jump process, in the low temperature phase below 204 K, the lattice can obviously no longer withstand the force imposed by the increasing electric field of the ordering molecular dipoles, so that the system relaxes by an increasing slip of the two sublattices against each other, thereby spring-loading the crystal. The difference in polarization between NMR and pyroeffect measurements (in C m^{-2})

$$\Delta pol = pol_{\text{NMR}} - pol_{\text{pyro}} \qquad (6.17)$$

enables to calculate the electric dipole moment (in D)

$$\mu_{\text{pol}} = \frac{\Delta pol \cdot V_{\text{ec}}}{3.36 \times 10^{-30} \text{Cm}} \qquad (6.18)$$

with the volume of the elementary cell, V_{ec}=177.6×10^{-30} m^3, leading finally to the sublattice displacement (in m)

$$x = \frac{\mu_{\text{pol}}}{q} \qquad (6.19)$$

with q=1.602×10^{-19} C. The full difference of 2.1 μC cm^{-2} between NMR (3-site) and pyroeffect measurements at 120 K is equivalent to a dipole moment of 1.11 D, which corresponds to displacement of 0.23 Å of the anion relative to the cation sublattice.

6-site Jump Model

The total difference in polarization in the low temeprature phase between the data from the analysis with the 6-site jump model and from the pyroeffect data is only slightly higher than that discussed above for the 3-site jump model. The value of 2.3 μC cm^{-2} in the low temperature limit corresponds to a displacement of 0.25 Å. In the intermediate temperature regime, the polarization curve obtained from the 6-site jump model is not identical to the pyroeffect data, but it shows the same trend. This is why we conclude that the total displacement in

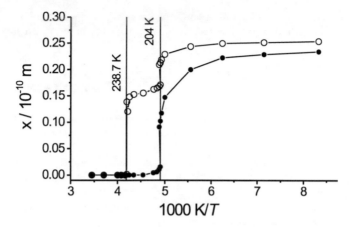

Figure 6.9: Lattice displacement obtained from the difference in polarization between the 3-site jump model (solid circles) and the 6-site jump model (open circles), and the macroscopic values. (For the polarization values see figure 6.8.)

case of the 6-site jump model is generated in a two step process, reaching 0.15 Å (1.4 μC cm^{-2}, 0.72 D) of displacement already in the intermediate and additional 0.10 Å (0.93 μC cm^{-2}, 0.48 D) in the low temperature phase. The displacement in the intermediate temperature phase is almost constant. The temperature dependence of the sublattice displacement is displayed in figure 6.9 for both models.

6.3.4   Critical Behavior

3-site Jump Model

Fitting the NMR polarization curve in the intermediate phase according to Landau's theory (see chapter 2.6) to a function

$$pol = pol_0 \left[1 - \left(\frac{T}{T_n} \right) \right]^c \qquad (6.20)$$

reproduces the data very well and results in a critical exponenct c=0.286 (T_n=237.30 K, pol_0=2.356 μC cm^{-2}). This is close to the value of 0.25 that

corresponds to a tricritical transition. It indicates a behavior at the border line
between a first (c undefined) and a second order (c=0.5) phase transition.

6-site Jump Model

Fitting the NMR data obtained with the 6-site jump model yields a critical exponent of c=0.156 (T_n=238.00 K, pol_0=3.998 μC cm^{-2}). This is much closer to a first
order than to a second order transition and already obvious from the polarization
plot (see fig. 6.8) as it shows a steplike change at the high temperature transition.

The macroscopic polarization data of Czarnecki et $al.$ is somewhat rounded off
just below the high temperature transition and was therefore not fitted. The
determination of a critical exponent around the low temperature transition in
the NMR curve was not possible due to the uncertainty of how to treat the
background from the intermediate phase.

6.4   Calorimetric Measurements

Heating rate dependent DSC measurements were performed on both,
PyClO$_4$-d$_5$ and PyBF$_4$-d$_5$ (see fig. 6.10). For PyClO$_4$-d$_5$, extrapolation to zero
heating rate revealed a supercooling by 1.5 K in case of the low and 4.4 K in
case of the high temperature transition that is compatible with the first order
transition described in literature.[17] In the case of PyBF$_4$-d$_5$ the extrapolated
phase transition temperature from the heating and the cooling curves agreed
within the experimental error of $\pm$0.3 K with each other so that no supercooling
was detected; but the low temperature phase transition was found to occur at
205.4 K instead of the reported value of 204 K, the high temperature transition
at 238.1 K instead of 238.7 K.[10,12] The DSC measurements performed with the
non-deuterated compounds PyBF$_4$ and PyClO$_4$ gave qualitatively identical results. Supercooling in the case of the perchlorate determined from extrapolation
to zero heating/cooling rate amounted to 1.2 K and 4.9 K for the low and the
high temperature transition, respectively. The phase transitions in the tetrafluoroborate were found at 205 K and at 238.5 K. The absolute temperature was not
calibrated for this experiment.

6.5   Thermodynamic Properties

For the analysis of thermal properties we start from the standard relation between
the enthalpy H and the internal energy U

$$dH = dU + p\,dV + V\,dp \qquad (6.21)$$

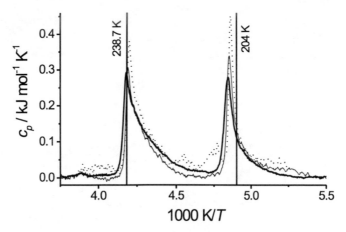

Figure 6.10: Base line corrected DSC measurements of $PyBF_4$-d_5 at 2 K min^{-1} (dotted line), 5 K min^{-1} (solid line) and 10 K min^{-1} (bold solid line) (all heating curves). Phase transitions are indicated with vertical solid lines (literature values, see ref. 10).

and note that

$$dU = dQ + dW_{vol} + dW_{fe} = dQ - pdV + dW_{fe} \qquad (6.22)$$

where the unconventional term W_{fe} denotes the ferroelectric work in the crystal lattice (see below). Under condition of constant pressure we obtain

$$\Delta H_{(T_1 \rightarrow T_2)} = Q_p + W_{fe} = \int_{T_1}^{T_2} c_p \, dT + \int_{T_1}^{T_2} c_{fe} \, dT, \qquad (6.23)$$

where c_{fe} is a formal analog of the heat capacity which is not detectable calorimetrically. As the temperature is lowered below 238.7 K, the energy of the system decreases by alignment of the molecular dipoles. A fraction of this energy Q_p is released as heat. It is mostly the one that relates to de-excitation of lattice vibrational modes and it can be measured in a calorimeter. In contrast, the other fraction, W_{fe}, is stored in the capacitance of the polarizing domains and as potential energy of the spring-loaded domains. Now it is possible to derive from the NMR data those thermodynamic properties of the system which relate to the orientational order of the cations.

3-site Jump Model

In analogy to the van't Hoff relation for the temperature dependence of a chemical equilibrium the fractional enthalpy change ΔH_n between two successive temperature points n and $n+1$ can be written as

$$\Delta H_n = -f \; R \; \frac{d(lnK)}{d(1/T)} = \Delta H^\circ, \tag{6.24}$$

where

$$f_{12} = \left(\frac{p_2 + p_2'}{p_1 + p_2 + p_2'}\right)_{n+1} - \left(\frac{p_2 + p_2'}{p_1 + p_2 + p_2'}\right)_n \tag{6.25}$$

$$f_{12} = (1 - p_1)_{n+1} - (1 - p_1)_n = p_{1,n} - p_{1,n+1} \tag{6.26}$$

is the fraction of molecules changing from p_1 to p_2 or p_2' within this temperature interval, and K is the equilibrium constant (see chapter 3.2.2). ΔH_n represents the difference in activation energies for a jump process out of and into the highest populated orientation. From ΔH_n we obtain the NMR-derived "heat capacities"

$$c = \frac{\Delta H_n}{\Delta T_n} = c_p + c_{fe}, \tag{6.27}$$

at intermediate temperatures

$$T = \frac{1}{2} \; (T_n + T_{n+1}). \tag{6.28}$$

They are shown in figure 6.11 together with the base line corrected conventional calorimetric data. Deviations of the phase transition temperatures to the literature values are due to the mean temperature values used in the analysis of the NMR data in combination with a minimum temperature difference of only 1 K in the measurements. Clearly, an analogous behavior is observed, but the NMR-derived spikes are much sharper, leading to higher values at the phase transition temperatures. More meaningful are perhaps the integrated enthalpy changes (see fig. 6.12)

$$\Sigma\Delta H_n = \Sigma(c \; \Delta T_n). \tag{6.29}$$

In the intermediate temperature range the NMR-derived enthalpy change amounts to 4.0 kJ mol^{-1}, which is only slightly more than the DSC value of 3.2 kJ mol^{-1}. The error in the difference is too large to permit a meaningful estimate of the energy that is stored in the capacitance of the polarized domains. More dramatic

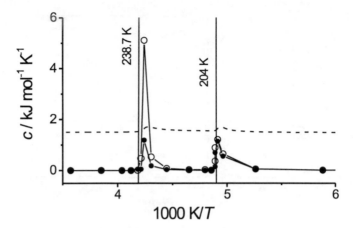

Figure 6.11: Heat capacity, $c = c_p + c_{fe}$, calculated from the population for the 3-site jump model (solid circles), for the 6-site jump model (open circles; for populations see fig. 6.2) and c_p from DSC measurements (10 K min^{-1}; see also fig. 6.10) (dashed line). DSC data are shifted upward by 1.5 kJ mol^{-1} K^{-1} to permit comparison.

is the situation in the low temperature phase. From NMR we derive an enthalpy change between 120 K and 204 K of 5.0 kJ mol^{-1}, while only 1.6 kJ mol^{-1} were measured calorimetrically. This means that most of the energy is stored as strain energy due to the relative displacement of the two sublattices. All results are listed in table 6.3.

6-site Jump Model

For the 6-site jump model the fractional enthalpy change ΔH_n consists of the individual enthalpy changes:

$$\Delta H_n = \Delta H_{12} + \Delta H_{23} + \Delta H_{34} \qquad (6.30)$$

which are related to the jump processes from $0°$ to $\pm 60°$, from $\pm 60°$ to $\pm 120°$ and from $\pm 120°$ to $180°$, respectively, and calculated according to equation 6.24. The individual rate constants K_{12}, K_{23} and K_{34} are given by the ratio of populations (see also chapter 3.2.2). The individual factors f amount to:

$$f_{12} = p_{1,n} - p_{1,n+1} \qquad (6.31)$$

$$f_{23} = f_1 - 2(p_{2,n+1} - p_{2,n}) \qquad (6.32)$$

$$f_{34} = p_{4,n+1} - p_{4,n} \qquad (6.33)$$

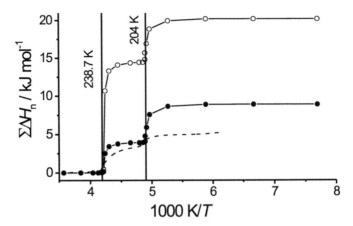

Figure 6.12: Total change of enthalpy from simulated NMR data for the 3-site jump model (solid circles), for the 6-site jump model (open circles) and from DSC measurements (10 K min^{-1}) (dashed line).

and give the fraction of molecules changing from p_1 to p_2 and p_2' (f_{12}), from p_2 and p_2' to p_3 and p_3' (f_{23}) and from p_3 and p_3' to p_4 and p_4' (f_{34}) within the temperature interval n and $n+1$. From the fractional enthalpy change ΔH_n the NMR-derived "heat capacities" are derived as described already for the 3-site jump model (see equation 6.27). They are also shown in figure 6.11. At the low temperature transition the values are almost identical to those derived from the 3-site jump model with respect to height and width. But at the high temperature transition the peak for the 6-site model is much higher. This is shown more clearly in figure 6.12 which gives the integrated enthalpy changes. Within the intermediate temperature phase the value for the integrated enthalpy change rises to 15 kJ mol^{-1} which is nearly five times more than derived by DSC and nearly four times more than determined with the 3-site jump model. In the low temperature regime there is observed an additional change of enthalpy of 5.5 kJ mol^{-1}. That is comparable to that found in the 3-site jump case, but once more much higher than found in DSC. All results are listed in table 6.3.

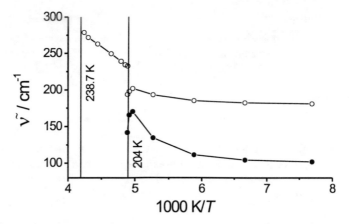

Figure 6.13: Frequency from simulated NMR data for the 3-site jump model (solid circles) and for the 6-site jump model (open circles). (For change of enthalpy see figure 6.12 and for sublattice displacement see figure 6.9.)

6.6   Harmonic Approximation

The difference between the NMR and DSC enthalpy changes

$$\Delta E_x = \Sigma \Delta H_n(\text{NMR}) - \Sigma \Delta H_n(\text{DSC}) \tag{6.34}$$

(see figure 6.12), maps the potential along the displacement coordinate x that is calculated from the polarization (see discussion of figure 6.8 and equation 6.15). On this basis we estimate the harmonic force constant k

$$\Delta E_x = \frac{1}{2} \, k \, x^2 \tag{6.35}$$

and the lattice vibrational frequency along this coordinate

$$\tilde{\nu} = \frac{\omega}{c} = \frac{1}{2\pi c} \sqrt{\frac{k}{\mu}} \tag{6.36}$$

with

$$\mu = \frac{1}{N_L} \frac{\text{M(PyH}^+)\text{M(BF}_4^-)}{(\text{M(PyH}^+) + \text{M(BF}_4^-))} = 6.917 \times 10^{-23}\text{g} \tag{6.37}$$

and $\text{M(PyH}^+) = 80.1$ g mol^{-1}; $\text{M(BF}_4^-) = 86.8$ g mol^{-1}.

3-site Jump Model

For the 3-site jump model we obtain $k = 26$ N m^{-1} and a frequency of 102 cm^{-1} at 120 K. The latter amounts to ca. 2/3 of the direct Raman spectroscopic determination by Ecolivet et al.[13]

6-site Jump Model

Calculating these values for the 6-site jump model yields $k = 82$ N m^{-1} and a frequency of 181 cm^{-1} at 120 K. The latter amounts to ca. 4/3 of the direct Raman spectroscopic determination by Ecolivet et al.[13] The frequency is almost constant in the low temperature phase as expected for a harmonic potential, but at the same time increasing from 230 cm^{-1} up to 280 cm^{-1} in the intermediate temperature regime.

Both simulation models yield frequency values which within the frame of several simplifying approximations are taken to be in very satisfactory qualitative agreement. But the 3-site model gives much lower, the 6-site model in contrast to that much higher values. Consequently, the determination of the frequency does not give information about the quality of the different simulation models.

6.7   Conclusion

The experimental ^{2}H NMR line shapes of PyBF$_4$-d$_5$ show a fast rotational motion for the pyridinium cation about a pseudo $C6$ axis in the high temperature phase. It is progressively impeded with decreasing temperature, resulting in a nearly static molecule at 120 K. The most distinct changes in motion take place near the phase transitions.

A simple 2-site jump model is able to simulate the quadrupole echo experiments, but it fails for the description of the T_{1Z} data and partially relaxed spectra. The present NMR data can adequately be described by both the 3-site and the 6-site jump model. It is therefore not possible to make a final statement about the correct model for the cation motion in the system examined here. The 3-fold rotation-inversion axis in the high temperature phase is compatible with a 6-fold jump process, which is also plausible based on the pseudo sixfold symmetry of the pyridinium ion. At both phase transitions the symmetry is lowered, but the structural change appears to be only marginal, suggesting that the 6-fold jump model may still be a better approximation and probably more plausible than the simpler threefold model. In order to demonstrate the extent of the model dependence of the resulting parameters we have investigated both models in detail.

The very good reproduction of the experimental spectra inlcuding the quadrupole echo and the inversion recovery experiments by both the 3-fold and the 6-fold jump model results in a simulated T_{1Z} curve which is identical to the experimental one. The minimum is found in the intermediate temperature phase showing that the dynamics of the pyridinium cations occurs at about the Lamor frequency of 46.07 MHz. The rate constant plot shows linear behavior in the high and in the low temperature phase resulting in activation energies of 19 kJ mol^{-1} and 10 kJ mol^{-1}, respectively. Only in the intermediate temperature phase several data points around the T_{1Z} minimum deviate from linearity. This behavior, and the unreasonably high activation energy and frequency factor, point to the presence of a composite process. If, for an alternative analysis, the high temperature frequency factor is assumed to remain valid also for the intermediate phase the experimental correlation times can be matched also with a stepwise increase of the activation energy from 19 to 24 kJ mol^{-1} over the intermediate phase. It reflects the way the energy offset between the different orientations is built up when the high temperature symmetry is broken and the pyridinium ions experience a preferential orientation in the electric field that is provided by the macroscopic polarization.

Enthalpy data derived from the NMR measurements reveal that for the low temperature phase only a fraction of the energy required to induce orientational disorder on heating is actually heat, the major fraction is released as potential energy. The same holds in the intermediate temperature regime for the 6-site jump model (but for the 3-site jump model the NMR-derived enthalpy change agrees nearly with the calorimetric determination). We postulate that in addition to cation ordering the classical ferroelectric mechanism contributes with an opposite effect on polarization, as the polarization obtained from the NMR data is much larger than the macroscopic one. An anion versus cation sublattice displacement of around 0.23-0.25 Å can account for the missing polarization. An estimate of the vibrational frequency in the low temperature phase yields values which are comparable with the experimental ones, and this supports our interpretation of the data on the basis of the sublattice displacement model.

Regarding the question of the order of the two phase transitions the summary is as follows. Typical criteria for first order are supercooling and phase coexistence. Other than for PyClO$_4$, which has almost the same lattice structure as PyBF$_4$ and two ferroelectric phases as well, but where there is no doubt that both transitions are of first order,[17] no evidence of supercooling is found when extrapolating to zero heating/cooling rates. Nevertheless, the spectra at 204 K, 205 K and 238 K reveal the contributions of two spectra with slightly different splittings, but it is not excluded that this could be due to temperature inhomogeneity. ALC-μSR measurements showed a steplike change of the linewidths for PyClO$_4$ but a smooth behavior for PyBF$_4$ at the high temperature transition.[149] The specific heat changes show very broad tails on the low temperature side of each transition as it is typical for order-disorder transitions. The two features sharpen only slightly upon reduction of the heating rate. The ferroelectric polarization curves are likewise discontinuous, but with a broad onset; only the high temperature transition in the 3-site jump model may show a continuous behavior. The latter fits very well critical behavior with a close to tricritical exponent, pointing to a border line case between first and second order, whereas the critical exponent determined for the 6-site jump process is closer to a first order transition. It has to be pointed out that both ferroelectric polarization mechanisms, the cation orientational order and the sublattice relative displacement, have a continuous onset and satisfy the Ehrenfest requirement for second order phase transitions. While the results presented here tend to support Czarnecki's interpretation of the phase transitions[17] they are clearly at variance with the work by Hanaya *et al.* who proposed that ferroelectricity originates from the orientational ordering of the tetrafluoroborate anions.[79] While it is plausible that the anion disorder contributes significantly to heat capacity and entropy it cannot contribute to ferroelectricity unless the anion is significantly distorted from tetrahedral symmetry. To us it is most likely that PyBF$_4$ is a border line case between first and second order.

It would be interesting to see the results of an up to date structure determination of the low temperature phase and the results of analogous NMR experiments with PyClO$_4$.

Table 6.3: NMR results compared to literature data.

	Method	lt [a]	it [b]	ht [c]	units
activation energy $E_A^{\neq}$	3-site [d]	10.0	46.4	19.3	kJ mol^{-1}
	6-site [d]	11.2	54.0	19.1	kJ mol^{-1}
frequency factor A [e]	3-site [d]	3.7×10^{10}	6.1×10^{19}	8.9×10^{13}	s^{-1}
	6-site [d]	1.8×10^{11}	1.2×10^{22}	3.3×10^{14}	s^{-1}
polarization difference Δpol	3-site [d]	2.2	1.4	0.0	μC cm^{-2}
	6-site [d]	0.9	2.9	0.0	μC cm^{-2}
	pyro [f]	0.1	1.4	0.0	μC cm^{-2}
critical exponent c	3-site [d]	-	0.286	-	
	6-site [d]	-	0.156	-	
sublattice displacement x	3-site [d]	0.23	0.0	0.0	Å
	6-site [d]	0.10	0.15	0.0	Å
enthalpy change $\Sigma\Delta H_n$	3-site [d]	5.0	4.0	0.0	kJ mol^{-1}
	6-site [d]	5.5	15	0.0	kJ mol^{-1}
	DSC	1.6	3.2	0.0	kJ mol^{-1}
frequency $\tilde{\nu}$	3-site [d]	102-170	-	-	cm^{-1}
	6-site [d]	181-200	230-280	-	cm^{-1}

[a]Low temperature phase; [b]intermediate temperature phase; [c]high temperature phase; [d] NMR; [e]frequency factor from axis intercept of simulated correlation time; [f]pyroeffect measurements by Czarnecki et al.[10]

Chapter 7

COMPARISON of μSR and NMR on PyBF$_4$

Muon Spin Resonance and the Deuteron Nuclear Magnetic Resonance were both used to investigate the dynamics in pyridinium tetrafluoroborate. Both techniques focus on the cation. μSR looks at the muonated aza-cyclohexadienyl radical ($^\bullet$C$_5$NH$_6$Mu$^+$) as the muon implanted into the sample adds as muonium atom to the unsaturated pyridinium ring. ^{2}H NMR only sees the perdeuterated pyridinium cation (C$_5$ND$_5$H$^+$). In μSR the dynamics of the radical are determined through the averaging hyperfine tensor. The ^{2}H NMR spectra refer to the quadrupolar tensor. The orientation of these tensors is different with respect to the investigated reorientational motion. Whereas the latter is axial and lies in the molecular plane of the cation parallel to the C–D bond, the former is non-axial and tilted about 25° out of the molecular plane. In fact, even the three different radicals (ortho-, meta- and para-radical) generated via Mu addition to the pyridinium cation when applying μSR already show different line width. This indicates that the radicals reveal only approximately the parent molecule dynamics. Nevertheless, both techniques seem to be appropriate to describe the investigated system, as the collective characteristics of the parent compound are reproduced qualitatively very well. This is nicely shown with PyClO$_4$. With the μSR technique it is obviously possible not only to detect the first order transition but also supercooling, which clearly was not observed in PyBF$_4$.

It has to be pointed out that with ALC-μSR meaningful spectra of PyBF$_4$ were only observable down to a temperature of 234 K. The very broad nature of the resonances below this temperature indicates that the correlation time for the motion corresponds approximately to the critical timescale of $\tau_{\text{ALC}} \approx 50$ ns. Consequently, a direct comparison of the data obtained from μSR and ^{2}H NMR has to be limited to the temperature range from 234 K to 290 K. Obviously this is a disadvantage of the μSR technique. Whereas it is possible to investigate both phase transitions in PyBF$_4$ with ^{2}H NMR, μSR is limited to the high temperature transition in the present context.

Qualitatively both techniques observe identical results with respect to the type of dynamics of the pyridinum cations in pyridinium tetrafluoroborate. The high temperature phase exhibits a fast rotational motion of the cation about the axis perpendicular to the molecular plane. This rotation is more and more hindered upon decreasing temperature with the most prominent changes taking place at the phase transition temperature.

In μSR the rotational motion is deduced from an asymmetric line shape of the Δ_1 resonances. The D_{zz} value is used to measure the extent of this asymmetry. For the aza-cyclohexadienyl radicals in muonated PyBF$_4$ it is obviously smaller than expected for the plain rotation of a cyclohexadienyl-type radical indicating additional motional averaging. This could be a tumbling or wobbling motion of the rotational axis. In ^{2}H NMR a fast wobbling motion of the axis would have decreased the experimental splitting of the Pake type pattern significantly. But this is not the case as the spectra in the high temperature phase show almost half of the splitting observed for the rigid case, a clear indication for a fast uniaxial rotation without additional motional averaging. It might be that the wobbling motion of the rotation axis takes place outside the time window of the ^{2}H NMR technique; but if the wobbling motion is slow, relaxation effects should have been observable.

Comparing the D_{zz} from μSR with the population probabilities obtained from the simulation (6-site jump model) in NMR shows that the trend with temperature is qualitatively identical. Both variables reveal the increasing motional hindering with decreasing temperature. Both are constant in the high temperature phase. At the high temperature phase transition the D_{zz} values change more smoothly, which is reminiscent of the continuous transition, whereas the change in population from NMR is more steplike and comparable to that in a discontinuous phase transition. As PyClO$_4$ was investigated with μSR as well it became obvious that ALC-μSR is able to distinguish between continuous and discontinuous transitions. For ^{2}H NMR this has not been shown yet. Investigations on PyClO$_4$ therefore would be very interesting as NMR on PyBF$_4$ revealed a more discontinuous change in population, which was not expected. Whereas μSR is only able to observe the motional hindering qualitatively, the population probabilities obtained from NMR enable us to quantify the increasing ordering of the electric dipole moments of the pyridinium cations.

However, there is no possibility to deduce information about the kinetics from the population data. These informations are obtained from the correlation time, which are received from the simulations of the partially relaxed NMR spectra. The correlation time gives the average time between two events, *i.e.*, between two jump processes in the present context, and therefore information about the rate of the cation rotation. It is not yet possible to deduce the rate constant of the rotational motion directly from the μSR data as the corresponding theoretical background is still missing. Only the dynamic broadening at temperatures below 234 K shows that the correlation time of the motion corresponds approximately to the critical ALC time scale here ($\tau_{\mathrm{ALC}} \approx 50$ ns; k $= 1/\tau_{\mathrm{ALC}} = 2 \times 10^7$ s^{-1}). This is surprising as values for the correlation time around the μSR time scale are observed in NMR only at temperatures of 160 K and 140 K with the 3-site and the 6-site jump model, respectively (see also figure 6.7). This means that μSR resonances should have been detectable down to this temperature range before disappearing due to dynamic line broadening. Thus, we conclude that additional relaxation processes take place in the high temperature phase. This emerged already from the line shapes at room temperature which were not exactly axial. Fitting of the resonances was often only possible by permitting an unlimited relaxation parameter.

Obviously both techniques deal with different time scales. This makes it difficult to perform a direct comparison of the data. But on the other hand applying both techniques renders complementary information. That is, μSR reveals a wobbling motion of the rotation axis in the high temperature phase whereas it is not able to detect the low temperature transition. Comparison with correlation times obtained from NMR indicates additional relaxation processes in the high temperature phase. NMR does not see the wobbling, but it is a very suitable technique to observe the cation dynamics in PyBF$_4$ in all three phases. In addition, the analysis of the NMR data permits the determination of thermodynamic and kinetic information quantitatively. Here, it has to be pointed out that using the NMR data for such purpose always depends strongly on the simulation model which is used. And as outlined earlier the consistency of experimental and simulated spectra only confirms but does not prove the mechanism underlying the model; it just cannot be excluded.

Chapter 8

SUMMARY

ALC-μSR was used to investigate interactions of cyclohexadienyl radicals with different transition metal cations accomodated in zeolites NaX and NaY. Therefore zeolite NaY was exchanged with Pt^{2+}, Pd^{2+}, Ag^+, Ni^{2+} and Zn^{2+} ions; NaX was exchanged with Mn^{2+}. To permit a direct comparison and to determine the metal cation influence the pure zeolites NaY and NaX were loaded with benzene and investigated with ALC-μSR as well. Radiating a sample containing unsaturated organic molecules like benzene leads to muon addition to the double or triple bond resulting in an organic radical, *i.e.*, the cyclohexadienyl radical in the present context.

Although zeolite NaY and NaX differ only in the Si/Al ratio and therefore in the number of extra framework cations the ALC-μSR spectra look quite different. The cyclohexadienyl radical in NaX is accomodated exclusively at the window sites not showing any sodium interaction at all. This is surprising, as the number of extra framework sodium cations is larger in X than in Y zeolite. The NaY structure shows three different types of radicals. According to Fleming *et al.* these radicals correspond to the window site and to an interaction with sodium cations at site SII, one in exo and one in endo configuration. The three corresponding methylene proton Δ_0 resonances were observed as well. Nevertheless, the spectra of C_6H_6/NaY are up to now not fully interpreted, as the assignment of an additional Δ_0 resonance of quite large intensity (G) is still unknown. In addition, two of the Δ_1 lines seem to be superimposed by additional resonances. The origin of the latter is still an open question. Due to this lack of information a direct comparison to the metal exchanged zeolite spectra was impossible. Nevertheless, an interpretation was possible. C_6H_6/MnNaX, C_6H_6/PdNaY and C_6H_6/PtNaY did not show a radical/metal ion interaction. For C_6H_6/MnNaX it is most likely that the radical electron interacts with the unpaired electrons of the paramagnetic manganese ion. This would result in relaxation processes which prevent the occurence of ALC-μSR resonances. Sample analysis of C_6H_6/PdNaY let us conlcude that redox processes during the sample preparation eliminated the benzene. Perhaps palladium did act as a catalyst here or even was involved in the redox reaction as a reactant. In the case of C_6H_6/PtNaY sample analysis revealed that the platinum ions were shielded by ammonia ligands. Therefore only benzene located at the window sites was observed. As all resonances in Pt-NaY are very weak it might also be possible that benzene was at least partially oxidized by the platinum. C_6H_6/AgNaY, C_6H_6/ZnNaY and C_6H_6/NiNaY indeed did show a radical/metal ion interaction. Comparison of the line position and the derived hfcs yields reasonable values compared to literature data. Calculating the spin densities shows that the values for the transition metal ions are all around $2.8\pm0.5\%$. These values are smaller than derived and calculated for the alkali metal ions ($\approx13\pm3\%$). This is surprising as the alkali metal ions are expected to show a weaker interaction with the radical and therefore smaller spin densi-ties. It has to be pointed out that all spectra showing transition metal/radical interactions reveal Δ_1 resonances of only very weak intensity. This indicates additional motional averaging which might be attributed to site hopping of the cyclohexadienyl radicals within the four tetrahedrally arranged sodium cations within the supercage. At the same time the metal Δ_0 resonance should decrease in intensity. Otherwise this would mean that the metal/radical sorption complex performs the site hopping, not only the cyclohexadienyl radical. The analysis of the ALC-μSR spectra of C_6D_6/NaY is still under way and supposed to deliver further information helpful for a correct interpretation.

The advantage of μSR in the present context is that it can handle the very low radical concentrations in the zeolite framework. Clearly the disadvantage is the sample preparation procedure in a closed system, $i.e.$, the metal cell. Sample analysis is only possible after the μSR measurements under destruction of the sample itself. And even then the conditions are not identical (pressure, contact to air).

Pyridinium tetrafluoroborate and pyridinium perchlorate were investigated by ALC-μSR with respect to the cation dynamics. Additional transverse field mea-surements were undertaken to permit the assignment of the resonances. Both compounds show three different phases. The high temperature transition of both salts is paraelectric-ferroelectric. In PyBF$_4$ this transition is continuous, whereas in the perchlorate it is discontinuous. Investigations of the dynamics of the pyri-dinium cations reveal information about the process how ferroelectricity is built up and how the order of a phase transition is determined. The method exclusively monitors the cation properties. Comparison with macroscopic data therefore per-mits the determination of the individual ionic contribution.

As expected PyBF$_4$ and PyClO$_4$ show identical ALC-μSR spectra. The spectra reveal three different radicals, the ortho- the meta- and the para-radical, upon muonium addition to the pyridinium cation. The ipso-adduct (muonium addition to the ring nitrogen) was not observed in ALC-μSR. The line shape of the muon Δ_1 resonances of all three radicals shows that the aza-cyclohexadienyl radicals perform fast rotation about the molecular plane normal. In addition, a wobbling or tumbling motion of the rotation axis in the high temperature phase was observed in both compounds. The line width of the ortho muon Δ_1 resonance in PyBF$_4$ decreases continuously when passing the high temperature transition as expected for second order. In clear contrast, the line width of the corresponding resonance in the perchlorate changes steplike reminiscent of the first order transition. Due to line broadening at lower temperatures in both compounds an investigation of the low temperature transition with ALC-μSR was not feasible.

Dynamic ^{2}H NMR spectroscopy employing line shape studies and spin-lattice relaxation experiments was used to investigate the molecular dynamics of the perdeuterated pyridinium cations in pyridinium tetrafluoroborate. Special interest lies in the two solid-solid second order phase transitions occurring at 204 K and 238.7 K. The experimental spectra show that the pyridinium cations perform fast rotation about the pseudo $C6$ axis at high temperatures. Decreasing the temperature leads to successive slowing down of the rotational motion resulting in nearly completely immobile and preferentially oriented pyridinium cations at low temperatures.

A simple 2-site model was able to interpret the quadrupole echo experiments but failed when simulating the inversion recovery measurements. Simulating the spectra with a 3-site or a 6-site jump model shows that both are appropriate to describe the experimental NMR data. Therefore it is not possible to distinguish which of them is the correct one. The 6-fold jump process was chosen with respect to the 3-fold rotation-inversion axis of the high temperature phase; but it is also plausible with respect to the pseudo sixfold symmetry of the pyridinium ions. The structural changes at the high temperature transition are minute. Therefore the 6-fold symmetry is expected to be still a good approximation. The 3-fold jump model was expected to deliver comparable good results in simulating the spectra. As the 3-fold symmetry is included in the high temperature symmetry of PyBF$_4$ it was considered in the investigations as well to demonstrate the model dependence of the resulting parameters.

The simulations obtained with both models show very good correspondence to the experimental spectra. The T_{1Z} curve reveals a motion of the pyridinium cation in the range of the Lamor frequency of 46.07 MHz. Activation energies derived from the rate constant plot in the high and the low temperature phase are comparable to the results for other pyridinium salts and even for benzene. Only in the intermediate phase deviations are observed which could be interpreted as a stepwise increase of the activation energy reflecting the way the energy offset between the different orientations is built up when the high temperature symmetry is broken. Both, 3-site and 6-site model, yield population probabilities which permit the calculation of the cation orientational contribution to the polarization and the enthalpy change. For both models distinct deviations from the macroscopic prop-

erties are found. We therefore suggest that in addition to the cation orientation
the classical ferroelectric mechanism operates which has a partly compensating
effect. A displacement of anion vs. cation sublattice by 0.23-0.25 Å can account
for the missing contribution to polarization. An estimate of the vibrational frequencies in the low temperature regime compared to experimental data verifies
this displacement model.

Both these ferroelectric mechanisms, the cation orientation and the classical ferroelectric mechanism, are potentially continuous in their onset, which is compatible
with the second order transitions in $PyBF_4$ proposed by Czarnecki et al.; but it
appears that they may both be border line cases between second and first order.
Line shape analysis of the ALC-μSR spectra indicate quite clearly a high temperature transition of second order. The broad tails at the low temperature side of
DSC measurements do so for both transitions as well. In addition, supercooling
was observed in DSC and ALC-μSR only for $PyClO_4$, but not for $PyBF_4$. In
contrast, the critical behavior for the ferroelectric polarization in $PyBF_4$ is closer
to first than to second order. The results presented here are clearly at variance
with Hanaya et al. suggesting the origin of ferroelectricity from the orientational ordering of the tetrafluoroborate anions. Analogous NMR measurements
on $PyClO_4$ would provide further information with respect to this point. In addition, it certainly would be helpful to have an up to date structure determination
of $PyBF_4$ in the low temperature phase to verify the interpretation presented in
the present context.

A direct comparison of the NMR and the μSR results obtained from the measurements on $PyBF_4$ is difficult. Clearly both methods deal with different time scales.
Thus, they render complementary results. Whereas ALC-μSR observes a wobbling motion of the rotation axis in the high temperature phase it is not possible
to investigate the low temperature transition due to dynamic line broadening. On
the other hand NMR does not see the wobbling motion of the rotation axis but is
otherwise a very good method to investigate of the cation dynamics in pyridinium
tetrafluoroborate within all three phases. ALC-μSR has shown to be sensitive to
the order of phase transitions, as the comparison of the data for the perchlorate and the tetrafluoroborate reveals. The perchlorate was not investigated with
^{2}H NMR so far. This would be very interesting as the population probabilities
obtained from NMR for the tetrafluoroborate show a rather steplike change at
the phase transition temperatures although this transition is known to be continuous. With the help of the simulation of the NMR spectra it is possible to
derive kinetic and thermodynamic information about the pyridinium cation. This
feature is very interesting as it enables a comparison with macroscopic data and
therefore the determination of the individual ionic contribution to those physical
properties. This was not yet possible with the μSR results as the corresponding
theoretical background is still missing. But is has to be pointed out that the data

determined from the NMR simulations always depend strongly on the underlying model. And as mentioned earlier the simulation models only confirm a proposed mechanism, they do not prove it. Only in the case that the experimental and the simulated spectra do not agree, the underlying model certainly has to be excluded.

Bibliography

[1] *Chem. Eng.* **108**, 59 (2001).

[2] J. Weitkamp; L. Puppe, *Catalysis and Zeolites*, (Springer-Verlag, 1998).

[3] I. E. Maxwell, *Adv. Catal.* **31**, 1 (1982).

[4] J. A. Rabo, *Zeolite Chemistry and Catalysis*, (ACS Monograph, 1976).

[5] D. W. Breck, *Zeolite Molecular Sieves*, (John Wiley, 1974).

[6] J. W. Wąsicki; W. Nawrocik; Z. Pająk; I. Nataniec; A. V. Belushkin, *Phys. Stat. Sol. A* **114**, 497 (1989).

[7] J. A. Ripmeester, *J. Chem. Phys.* **85**, 747 (1986).

[8] J. A. Ripmeester, *Can. J. Chem.* **54**, 3453 (1976).

[9] S. Lewicki; J. Wąsicki; P. Czarnecki; I. Szafraniak; A. Kozak; Z. Pająk, *Mol. Phys.* **94**, 973 (1998).

[10] P. Czarnecki; A. Katrusiak; I. Szafraniak; J. Wąsicki, *Phys. Rev. B* **57**, 3326 (1998).

[11] P. Czarnecki; W. Nawrocik; Z. Pająk; J. Wąsicki, *Phys. Rev. B* **49**, 1511 (1994).

[12] I. Szafraniak; P. Czarnecki; P. U. Mayr, *J. Phys.: Condensed Matter* **12**, 643 (2000).

[13] C. Ecolivet; P. Czarnecki; J. Wąsicki; S. Beaufils; A. Girard; L. Bobrowicz-Sarga, *J. Phys.: Condensed Matter* **13**, 6563 (2001).

[14] L. Bobrowicz-Sarga; P. Czarnecki; S. Lewicki; I. Nataniec; J. Wąsicki, *Materials Science Forum* **321**, 1107 (2000).

[15] P. Czarnecki; W. Nawrocik; Z. Pająk; J. Wąsicki, *J. Phys.: Condensed Matter* **6**, 4955 (1994).

[16] P. Czarnecki; J. Wąsicki; Z. Pająk; R. Goc; H. Małuszyńska; S. Habryło, *J. Mol. Struct.* **404**, 175 (1997).

[17] J. Wąsicki; S. Lewiki; P. Czarnecki; C. Ecolivet; Z. Pająk, *Mol. Phys.* **98**, 643 (2000).

[18] Z. Pająk; P. Czarnecki; J. Wąsicki; W. Nawrocik, *J. Chem. Phys.* **109**, 6420 (1998).

[19] P. Czarnecki; H. Małuszyńska, *J. Phys.: Condensed Matter* **12**, 1 (2000).

[20] J. Wąsicki; P. Czarnecki; Z. Pająk; W. Nawrocik; W. Szczepanski, *J. Chem. Phys.* **107**, 576 (1997).

[21] A. Kozak; J. Wąsicki; Z. Pająk, *Phase Trans.* **57**, 153 (1996).

[22] A. Kozak; M. Grottel; J. Wąsicki; Z. Pająk, *Phys. Stat. Sol.* **141**, 345 (1994).

[23] J. W. Wąsicki; A. Kozak; Z. Pająk; P. Czarnecki; A. V. Belushkin; M. A. Adams, *J. Chem. Phys.* **105**, 9470 (1996).

[24] I. Szafraniak; P. Czarnecki; P. U. Mayr; W. Dollhoff, *Phys. Stat. Sol.* **213**, 15 (1999).

[25] Z. Pająk; P. Czarnecki; H. Małuszyńska; B. Szafrańska; M. Szafran, *J. Chem. Phys.* **113**, 848 (2000).

[26] M. Alexe; C. Harnagea; D. Hesse, *Physikalische Blätter* **56**, 47 (2000).

[27] A. Pignolet; M. Alexe, *Vacuum Solutions* (2000).

[28] R. G. Barnes, *Deuteron Quadrupole Coupling Tensors in Solids*, p. 335, (J. A. S. Smith, 1974).

[29] T. M. Alam; G. P. Drobny, *Chem. Rev.* **91**, 1545 (1991).

[30] H. W. Spiess, *Adv. Polym. Sci.* **66**, 23 (1985).

[31] E. Roduner, *Chem. Soc. Rev.* **22**, 337 (1993).

[32] E. Roduner, *Lecture Notes in Chemistry*, volume 49, p. 1, (G. Berthier; M. J. S. Dewar; H. Fischer; K. Fukui; G. G. Hall; J. Hinze; H. H. Jaffé; J. Jortner; W. Kutzelnigg; K. Ruedenberg; J. Tomasi, 1988).

[33] M. J. Ramos, *J. Chim. Phys.* **84**, 619 (1987).

[34] G. Schatz; A. Weidinger, *Nukleare Festkörperphysik*, (Teubner Studienbücher, 1992).

[35] E. Roduner, *Appl. Magn. Reson.* **13**, 1 (1997).

[36] E. Roduner, *Hyperfine Int.* **65**, 857 (1990).

[37] J. W. Schneider, *Avoided Level Crossing: A New Technique in Muon Spin Rotation to Study the Nuclear Hyperfine Structure of Muonium Centres in Semiconductors*, Ph.D. thesis, University of Zürich, 1989.

[38] R. F. Kiefl; S. R. Kreitzmann, *Perspective of Meson Science*, volume 1, p. 265, (Y. Yamazaki; K. Nakai; K. Nagamine, 1992).

[39] D. G. Fleming; M. Y. Shelley; D. J. Arseneau; M. Senba; J. J. Pan; E. Roduner, *J. Phys. Chem. B* **106**, 6395 (2002).

[40] S. H. Neddermeyer; C. D. Anderson, *Phys. Rev.* **54**, 88 (1938).

[41] J. I. Friedman; V. L. Telegdi, *Phys. Rev.* **105**, 1681 (1957).

[42] A. M. Brodskii, *Soviet Phys. JETP* **17**, 1085 (1963).

[43] E. Roduner; P. W. Percival; D. G. Fleming; J. Hochmann; H. Fischer, *Chem. Phys. Letters* **57**, 37 (1978).

[44] R. L. Garwin; L. M. Lederman; M. Weinrich, *Phys. Rev.* **105**, 1415 (1957).

[45] R. A. Swanson, *Phys. Rev.* **112**, 580 (1958).

[46] V. S. Evseev, *Muon Physics*, volume 3, p. 235, (V. W. Hughes, 1975).

[47] T. Yamazaki; K. Nagamine; O. Hashimoto; K. Sugimoto; K. Nakai; S. Kobaiashi, *Physica Scripta* **11**, 133 (1975).

[48] K. Nishiyama; K. Nagamine; H. Kitazawa; E. Torikai; H. Kohjima; I. Tanaka, *Hyp. Int.* **65**, 1015 (1990).

[49] E. Roduner; M. Stolmár; H. Dilger; I. D. Reid, *J. Chem. Phys. A* **102**, 7591 (1998).

[50] E. Roduner; G. Brinkman; P. W. L. Louwrier, *Chem. Phys.* **88**, 143 (1984).

[51] M. Schwager; H. Dilger; E. Roduner; I. D. Reid; P. W. Percival; A. Baiker, *Chem. Phys.* **189**, 697 (1994).

[52] M. Stolmár; E. Roduner, *J. Am. Chem. Soc.* **120**, 583 (1998).

[53] R. G. Griffin, *Methods Enzymol.* **72**, 108 (1981).

[54] J. H. Davis, *Biochim. Biophys. Acta* **737**, 117 (1983).

[55] K. Müller; P. Meier; G. Kothe, *Prog. Nucl. Magn. Reson. Spectrosc.* **17**, 211 (1985).

[56] R. R. Vold; R. L. Vold, *Adv. Magn. Opt. Res.* **16**, 85 (1991).

[57] J. Ripmeester, *Inclusion Compounds*, volume 5, p. 37, (J. L. Atwood; J. E. Davies; D. D. MacNicol, 1991).

[58] M. S. Greenfield; R. L. Vold; R. R. Vold, *Mol. Phys.* **66**, 269 (1989).

[59] E. Meirovitch; T. Krant; S. Vega, *J. Phys. Chem.* **87**, 1390 (1983).

[60] R. Poupko; E. Furman; K. Müller; Z. Luz, *J. Phys. Chem.* **95**, 407 (1991).

[61] D. A. Torchia; A. Szabo, *J. Magn. Res.* **49**, 107 (1982).

[62] D. S. Coombs, *The Canadian Mineralogist* **35**, 1571 (1997).

[63] A. F. Crønstedt, *Akad. Handl. Stockholm* **18**, 120 (1756).

[64] A. Dyer, *Introduction to Zeolite Molecular Sieves*, (John Wiley, 1988).

[65] A. N. Fitch; H. Jobic; A. Renouprez, *J. Phys. Chem.* **90**, 1311 (1986).

[66] *American Mineralogist* **39**, 92 (1954).

[67] R. L. Portsmouth; M. J. Duer; L. F. Gladden, *J. Chem. Soc.: Faraday Trans. 1* **91**, 559 (1995).

[68] G. Vitale; L. M. Bull; R. E. Morris; A. K. Cheetham; B. H. Toby; C. G. Coe; J. E. McDougall, *J. Phys. Chem.* **99**, 16087 (1995).

[69] L. M. Bull; N. J. Henson; A. K. Cheetham; J. M. Newsam; S. J. Heyes, *J. Phys. Chem.* **97**, 11776 (1993).

[70] J. Caro; H. Jobic; M. Bülow; J. Kärger; B. Zibrowius, *Adv. Catal.* **39**, 351 (1993).

[71] V. Voss; B. Boddenberg, *Surf. Sci.* **298**, 241 (1993).

[72] B. Zibrowius; J. Caro; H. Pfeifer, *J. Chem. Soc.: Faraday Trans. 1* **84**, 2347 (1988).

[73] H. Jobic; M. Bée; A. J. Dianoux, *J. Chem. Soc.: Faraday Trans. 1* **85**, 2525 (1989).

[74] R. Q. Snurr; A. T. Bell; D. N. Theodorou, *J. Phys. Chem.* **98**, 5111 (1994).

[75] H. Jobic; A. Renouprez; A. N. Fitch; H. J. Lauter, *J. Chem. Soc.: Faraday Trans. 1* **83**, 3199 (1987).

[76] A. J. Renouprez; H. Jobic; R. C. Oberthur, *Zeolites* **5**, 222 (1985).

[77] R. M. Macrae; B. C. Webster, *Physica B* (2003).

[78] P. M. M. Blauwhoff; J. W. Gosselink; E. P. Kieffer; S. T. Sie; W. H. J. Storck, *Catalysis and Zeolites*, volume 1, p. 437, (J. Weitkamp; L. Puppe, 1999).

[79] M. Hanaya; H. Shibazaki; M. Oguni; T. Nemoto; Y. Ohashi, *J. Phys. Chem. Sol.* **61**, 651 (2000).

[80] F. Giesselmann, *Selforganization in chiral liquid crystals*, p. 7, (W. Kuszynski, 1997).

[81] P. W. Atkins, *Physikalische Chemie*, (Wiley-VCH, 2001).

[82] P. W. R. Bessonette; M. A. White, *J. Chem. Ed.* **2**, 220 (1999).

[83] St. Elliott, *The Physics and Chemistry of Solids*, (Wiley, 1998).

[84] E. Roduner; I. D. Reid, *Isr. J. Chem.* **29**, 3 (1989).

[85] J. A. Weil; J. R. Bolton; J. E. Wertz, *Electron Paramagnetic Resonance*, (John Wiley & Sons, 1994).

[86] F. James; M. Roos, *Comput. Phys. Commun.* **10**, 343 (1975).

[87] E. Roduner; I. D. Reid; R. De Renzi; M. Riccó, *Ber. Bunsen-Ges. Phys. Chem.* **93**, 1194 (1989).

[88] N. J. Heaton; R. R. Vold; R. L. Vold, *J. Magn. Res.* **77**, 572 (1988).

[89] J. Schmider; K. Müller, *J. Phys. Chem. A* **102**, 1181 (1998).

[90] B. T. Smith; J. M. Bole; B. S. Garbow; Y. Ikebe; V. C. Klema; C. B. Moler, *Matrix Eigensystem Routines-EISPACK Guide*, (Springer-Verlag, 1976).

[91] R. J. Wittebort; E. T. Olejniczak; R. G. Griffin, *J. Chem. Phys.* **86**, 5411 (1987).

[92] R. R. Vold, *NMR Probes of Molecular Dynamics*, p. 27, (R. Tycko, 1994).

[93] H. W. Spiess; H. Sillescu, *J. Magn. Res.* **42**, 381 (1981).

[94] A. J. Vega; Z. Luz, *J. Chem. Phys.* **86**, 1803 (1987).

[95] R. J. Wittebort; A. Szabo, *J. Chem. Phys.* **69**, 1722 (1978).

[96] Y. Kim; A. N. Kim; Y. W. Han; K. Seff, *Proceedings of the 12th International Zeolite Conference*, volume 4, p. 2839, (M. M. J. Treacy; B. K. Marcus; M. E. Bisher; J. B. Higgins, 1999).

[97] S. M. Auerbach; N. J. Henson; A. K. Cheetham; H. I. Metiu, *J. Phys. Chem.* **99**, 10600 (1995).

[98] D. G. Fleming; D. J. Arseneau; J. J. Pan; M. Y. Shelley; M. Senba; P. W. Percival, *Appl. Mag. Res.* **13**, 1 (1997).

[99] P. W. Percival; R. F. Kiefl; S. R. Kreitzman; D. M. Garner; S. F. J. Cox; G. M. Luke; J. H. Brewer; K. Nishiyama; K. Venkateswaran, *Chem. Phys. Letters* **133**, 465 (1987).

[100] D. Yu; P. W. Percival; J.-C. Brodovitch; S.-K. Leung; R. F. Kiefl; K. Venkateswaran; S. F. J. Cox, *Chem. Phys.* **142**, 229 (1990).

[101] M. Stolmár, *Appl. Magn. Reson.* **13**, 173 (1997).

[102] B. Webster; R. M. Macrae, *Physica B* **289**, 598 (2000).

[103] L. Kevan, *Acc. Chem. Res.* **20**, 1 (1987).

[104] M. W. Anderson; L. Kevan, *J. Phys. Chem.* **91**, 4174 (1987).

[105] T. H. Bennur; D. Srinivas; P. Ratnasamy, *Microporous and Mesoporous Materials* **48**, 111 (2001).

[106] M. Stolmár; E. Roduner; H. Dilger; U. Himmer; M. Shelley; I. D. Reid, *Hyperfine Interact.* **106**, 51 (1997).

[107] D. R. Gee; J. K. S. Wan, *Can. J. Chem.* **49**, 160 (1971).

[108] C. Lamberti; M. Milanesio; C. Prestipino; S. Bordiga; A. N. Fitch; G. L. Marra, *ESRF Newsletter, Experiments Report* (2001).

[109] T. Sun; K. Seff, *Chem. Rev.* **94**, 857 (1994).

[110] L. R. Gellens; W. J. Mortier; J. B. Uytterhoeven, *Zeolites* **1**, 85 (1981).

[111] P. L. W. Tregenna-Piggott; E. Roduner; S. Santos, *Chem. Phys.* **203**, 317 (1996).

[112] M. V. Frash; R. A. van Santen, *Phys. Chem. Chem. Phys.* **2**, 1085 (2000).

[113] J. A. Lercher J. Penzien; Th. E. Müller, *Microporous and Mesoporous Materials* **48**, 285 (2001).

[114] M. A. Wassel; E. A. Sultan; F. M. Tawfik, *Asian J. Chem.* **4**, 891 (1992).

[115] M. Ziolek; H. G. Karge; W. Niešen, *Zeolites* **10**, 662 (1990).

[116] B. Boddenberg; A. Seidel, *J. Chem. Soc.: Faraday Trans.* **90**, 1345 (1994).

[117] K. Otsuka; J. Manda; A. Morikawa, *J. Chem. Soc.: Faraday Trans. 1* **77**, 2429 (1981).

[118] T. A. Egerton; F. S. Stone, *J. Chem. Soc.: Faraday Trans. 1* **69**, 22 (1973).

[119] R. A. Dalla-Betta; M. Boudart, *Proceedings of the 5th International Congress on Catalysis*, volume 1, p. 1329, (H. Hightower, 1973).

[120] S. T. Homeyer; Z. Karpiński; W. H. M Sachtler, *Recl. Trav. Chim. Pays-Bas* **109**, 81 (1990).

[121] V. Romannikov; K. Ione; L. Pedersen, *J. Catal.* **66**, 121 (1980).

[122] H. Du; R. Klemt; F. Schell; J. Weitkamp; E. Roduner, *Proceedings of the 12th International Zeolite Conference*, volume 4, p. 2665, (M. M. J. Treacy; B. K. Marcus; M. E. Bisher; J. B. Higgins, 1999).

[123] A. L. Lapidus; K. M. Minachev, *Neftehimiya* **18**, 212 (1978).

[124] A. K. Ghosh; L. Kevan, *J. Phys. Chem.* **94**, 1953 (1990).

[125] S. T. Homeyer; W. H. M Sachtler, *J. Catal.* **117**, 91 (1989).

[126] M. S. Tzou; B. K. Teo; W. M. H Sachtler, *J. Catal.* **113**, 220 (1988).

[127] G. Bergeret; T. M. Tri; P. Gallezot, *J. Phys. Chem.* **87**, 1160 (1983).

[128] G. Bergeret; P. Gallezot; B. Imelik, *J. Phys. Chem.* **85**, 411 (1981).

[129] G. A. Somorjai, *J. Phys. Chem.* **94**, 1013 (1990).

[130] S. Ciccariello; A. Benedetti; F. Pinna; G. Strukul; W. Juszczyk; H. Brumberger, *Phys. Chem. Chem. Phys.* **1**, 367 (1999).

[131] P. Gallezot; A. Alarcon-Diaz; I. A. Dalmon; A. J. Renouprez; B. Imelik, *J. Catal.* **39**, 334 (1975).

[132] A. K. Ghosh; L. Kevan, *J. Phys. Chem.* **94**, 3117 (1990).

[133] V. Yu. Borokov; V. A. Kaverinsky; V. B. Kazansky, *Proceedings of the International Symposium on Related Heterogeneous and Homogeneous Catalytic Phenomenon*, volume 1, p. 253, (B. Delmon; G. Jannes, 1975).

[134] G. Wendt; J. Finister; R. Schollner; H. Siegel, *Stud. Surf. Sci. Catal.* **7**, 978 (1981).

[135] G. Wendt; W. Morke; R. Schollner; H. Siegel, *Z. Anorg. Allg. Chem.* **488**, 197 (1982).

[136] T. Yashima; Y. Ushida; M. Ebisawa; N. Hara, *J. Catal.* **36**, 320 (1975).

[137] L. Bonneviot; D. Olivier; M. Che, *J. Mol. Catal.* **21**, 415 (1983).

[138] D. G. Fleming, *personal communication* (2001).

[139] S. M. Auerbach; L. M. Bull; N. J. Henson; H. I. Metiu; A. K. Cheetham, *J. Phys. Chem.* **100**, 5923 (1996).

[140] D. C. Doetschman; D. C. Gilbert; D. W. Dwyer, *Chem. Phys.* **256**, 37 (2000).

[141] Ch. J. Rhodes; H. Morris; I. D. Reid, *Magn. Reson. Chem.* **39**, 438 (2001).

[142] J. C. Ronfard Haret; A. Lablache Combier; C. Chachaty, *J. Phys. Chem.* **78**, 899 (1974).

[143] B. Beck; J. A. Villanueva-Garibay; K. Müller; E. Roduner, *Chem. Mater.* (2003).

[144] J. H. Ok; R. R. Vold; R. L. Vold; M. C. Etter, *J. Phys. Chem.* **93**, 7618 (1989).

[145] F. A. Bovey; E. W. Anderson; F. P. Hood; R. L. Kornegay, *J. Chem. Phys.* **40**, 3099 (1964).

[146] J. E. Anderson, *J. Magn. Res.* **11**, 398 (1973).

[147] J. Wąsicki; Z. Pająk; A. Kozak, *Z. Naturforsch.* **45**, (1990).

[148] T. Gullion; M. S. Conradi, *Phys. Rev. B* **32**, 7076 (1985).

[149] B. Beck; E. Roduner; H. Dilger; P. Czarnecki; D. G. Fleming; I. D. Reid; Ch. J. Rhodes, *Physica B* **289**, 607 (2000).

APPENDIX

Abbreviations

α-cage	supercage
β-cage	sodalite cage
Δ_0	resonance due to a muon-proton spin flip-flop
Δ_1	resonance due to a muon spin flip
Δ_2	resonance due to a muon-proton spin flip-flip
μSR	muon spin resonance
A, B, C ...	labeling for resonances not assigned yet
A	zeolite built from sodalite cages connected via double-sixrings
A^{m+}	m+ charged extra framework cations
AES	atom emission spectroscopy
AgNaY	sodium form of zeolite Y, partly exchanged with silver ions
ALC	avoided level crossing
BEA	zeolite Beta
BF_4^-	tetrafluoroborate anion
ClO_4^-	perchlorate anion
D6R	double six ring, hexagonal prism
DSC	differential scanning calorimetry
EA	elementary analysis
ec	elementary cell
ENDOR	electron nuclear double resonance
EPR	electron paramagnetic resonance
ESRF	European Synchrotron Radiation Facility
EXAFS	extended X-ray adsorption fine structure
FAU	zeolite faujasite
FID	free induction decay
FWHM	full width at half maximum

hfc	hyperfine coupling constant
HWHM	half width at half maximum
ICP	inductive coupled plasma
IR	infrared spectroscopy
LSX	low-silica zeolite X
m	meta
MnNaX	sodium form of zeolite X, partly exchanged with manganese ions
NaY	sodium form or zeolite Y
NaX	sodium form or zeolite X
NiNaY	sodium form of zeolite Y, partly exchanged with nickel ions
NMR	nuclear magnetic resonance
o	ortho
p	para
PdNaY	sodium form of zeolite Y, partly exchanged with palladium ions
PSI	Paul Scherrer Institute
PtNaY	sodium form of zeolite Y, partly exchanged with platinum ions
PVP	polyvinylpyridine
Py	pyridine
$PyBF_4$	pyridinium tetrafluoroborate, $C_5NH_6BF_4$
$PyBF_4$-d_5	perdeuterated pyridinium tetrafluoroborate, $C_5ND_5HBF_4$
$PyClO_4$	pyridinium perchlorate, $C_5NH_6ClO_4$
$PyClO_4$-d_5	perdeuterated pyridinium perchlorate, $C_5ND_5HClO_4$
PyH^+	pyridinium cation, $C_5NH_6^+$
pyro	pyroelectric effect measurements
RFA	X-ray fluorescense analysis
SAXS	small angle X-ray scattering
sc	supercage, α-cage
TEM	transmission electron microscopy
TF	transverse field
USY	ultrastable zeolite Y

X	synthethic zeolite of faujasite type, x/y=1.0-1.5
XPS	x-ray photoelectron spectroscopy
XRPD	X-ray powder diffraction
Y	synthethic zeolite of faujasite type, x/y=1.5-3.0
ZnNaY	sodium form of zeolite Y, partly exchanged with zinc ions
ZSM-5	zeolite ZSM-5 with a threedimensional channel system

Variables

γ	gyromagnetic ratio
$\Delta\nu$	splitting between singularities in ^{2}H NMR
η	order parameter
μ	electric dipole moment
μ	reduced mass
μ^+	positive muon
μ^-	negative muon
μ	magnetic moment
ν	neutrino
$\tilde{\nu}$	lattice vibrational frequency
π^+	positive pion
π^-	negative pion
ρ_s	spin population of the s-orbital
τ_1, τ_2	delay times
τ	correlation time
τ_c	correlation time in NMR
A	frequency factor
A	complex matrix
A_μ	muon hyperfine coupling constant
A'_μ	reduced muon hyperfine coupling constant $(A'_\mu=A_\mu/3.184)$
A_n	nuclear hyperfine coupling constant
a_o	tabulated isotropic hyperfine coupling constant
A_p	proton hyperfine coupling constant
As	Asymmetry in ALC
B_{res}	resonant magnetic field position
c	velocity of light
c	critical exponent
c	heat capacity: $c=c_p+c_v$

c_p	heat capacity at constant pressure
c_v	heat capacity at constant volume
$D_\perp$	perpendicular component of D_{zz}
$D_\parallel$	parallel component of D_{zz}
dip	resonance intensity
D_{zz}	anisotropy
e	electron
e^+	positron
$E_A^{\neq}$	activation energy of transition state
f	fraction of molecules changing from one orientation to another
I	spin
J	spectral density
k	rate constant
k	harmonic force constant
K	kinetic matrix
m	mass
m+	cation charge
M	molar mass
M_s	electron spin
n	number of moles
n+	cation charge
N_L	Avogadro's number
p	proton
p_1, p_2, p_3	population probability of orientations
pol	polarization
P_s	spontaneous polarization
q	elementary charge
Q	heat
$\dot{Q}$	heat flux
r	heating rate
R	gas constant (8.31441 J mol^{-1} K^{-1})

t	time
T	temperature
T_{1Z}	spin-lattice relaxation time
T_2	spin-spin relaxation time
V_{ec}	volume of the elementary cell
W	line width
x	sublattice displacement
x	number of moles of silicon
x/y	molar silicon to aluminium ratio
y	number of moles of aluminium

Indexing

$\neq$	transition state
$\perp$	perpendicular
$\parallel$	parallel
μ	muon
A	activation
d	deuterium
e	electron
ec	elementary cell
ir	inversion recovery
iso	isotropic
m	meta
n	nucleus
o	ortho
p	para
p	proton
p	constant pressure
qe	quadrupole echo
res	resonance/resonant
s	s-orbital
s	spontaneous
v	constant volume
x	lattice displacement

List of Figures

List of Tables

ZUSAMMENFASSUNG

Motivation

Zeolithe sind poröse, kristalline Aluminosilikatgerüste, die eine definierte Hohlraumstruktur aufweisen. Ihre negative Gerüstladung wird durch Kationen kompensiert, die nicht in das Gerüst eingebaut und daher leicht austauschbar sind. Zeolithe finden in Wasch- und Reinigungsmitteln sowie in Trennprozessen zahlreich Anwendung. Aber auch in der Katalyse sind sie von großer Bedeutung, nicht nur, aber besonders in der petrochemischen Industrie, in der sie unter anderem zur Herstellung von Kraftstoffen aus Rohöl eingesetzt werden. Insbesondere Zeolithe, deren negative Gerüstladung durch Übergangsmetallkationen kompensiert wird, sind hier von großem Interesse. Diese Metallkationen können durch zumeist einfache Ionenaustauschprozesse in den Zeolithen eingebracht werden. Abgesehen von der Hohlraumstruktur des Zeolithen selber sind für die Selektivität und die Effektivität des Katalysators insbesondere die Verteilung der Kationen im Zeolithen, ihr Aufenthaltsort in den Hohlräumen und damit die Zugänglichkeit für Adsorbatmoleküle sowie ihre Oxidationsstufe ausschlaggebend. Da es sich bei den Kationen um die aktiven Zentren des Katalysators handelt, sind auch die translatorische Mobilität von Edukten, Produkten und auftretenden Zwischenstufen, ihre Wechselwirkung mit den aktiven Zentren sowie ihre Reorientierungsdynamik von großem Interesse. In der Literatur werden zumeist Untersuchungen von Wechselwirkungen von paramagnetischen Metallen mit organischen Molekülen mittels ESR beschrieben. Der entgegengesetzte Fall von diamagnetischen Metallzentren, die mit Radikalen wechselwirken, findet weniger Berücksichtigung, obwohl angenommen wird, dass Radikale eine entscheidende Rolle als Zwischenstufe in katalytischen Prozessen spielen.

Die durch die Beladung mit Benzol erhaltenen Zeolithe stellen typische Wirts-Gast-Systeme dar, an denen bereits unzählige Untersuchungen zur Translations- und Reorientierungsdynamik vorgenommen worden sind. Dabei geht es in den meisten Fällen um die Dynamik der Gastmoleküle im Wirtsgitter. Eine derartige Dynamik in Festkörpern kennt man jedoch nicht nur von Wirts-Gast-Systemen. So weiss man beispielsweise von Molekülkristallen, dass deren Bausteine sehr wohl ungehinderte Rotationsbewegungen durchführen können. Voraussetzung ist, dass die Moleküle eine genügend hohe Symmetrie aufweisen, welche eine sterische Hinderung minimal hält. Dies ist insofern verblüffend, als dass ein Kristall in makroskopisch perfekter Ordnung aufgebaut ist und nicht auseinanderfällt, obwohl seine Bestandteile ungehindert rotieren. Etliche Beispiele für derartige Molekülkristalle findet man in der großen Familie der Pyridiniumsalze. Darunter befinden sich Verbindungen des Pyridinium-Kations sowohl mit einfachen (Cl^-, Br^-, I^-) als auch mit hochsymmetrischen (BF_4^-, ClO_4^-, IO_4^-, ReO_4^-, PF_6^-, SbF_6^-) Anionen. Alle Salze zeigen mindestens einen fest-fest Phasenübergang, oberhalb dessen sowohl die Kationen als auch die Anionen frei rotieren. Die Hochtemperaturphase ist in jedem Fall paraelektrisch. Auch unterhalb des Phasenüberganges bleibt eine gewisse Dynamik erhalten. Die Salze mit den tetraedrisch aufgebauten Anionen sowie auch das Fluoro-sulfonat werden unterhalb des Hochtemperaturüberganges ferroelektrisch. Die ferroelektrische Ordnung geht einher mit einer bevorzugt parallelen Ausrichtung der elektrischen Dipolmomente der Pyridiniumionen. Noch ist unklar, welche Faktoren neben der Symmetrieverwandtschaft der beteiligten Phasen dafür verantwortlich sind, dass viele dieser Phasenübergänge kontinuierlich, also zweiter Ordnung sind, während unter struturell ähnlichen Voraussetzungen andere Pyridiniumsalze sich nach erster Ordnung umwandeln. Außerdem ist auch unklar, wodurch derartige Übergänge generell ausgelöst werden. Um Phasenübergänge beschreiben und verstehen zu können, sind detaillierte Informationen über die Dynamik der Ionen und die Zeitskala dieser Dynamik notwendig. Außerdem sind die Korrelation zwischen der ferro-elektrischen Polarisation und der Dynamik genauso wie die Auswirkung sterischer Verhältnisse auf die Dynamik von Bedeutung, um die unterschiedlichen Eigen-schaften der Pyridiniumsalze erklären zu können. Ferroelektrische Materialien werden wegen ihres potentiellen Einsatzes in elektronischen Speichermedien oder in Sensoren intensiv untersucht; sie sind aber auch von prinzipiellem Interesse in Bezug auf das Verständnis intermolekularer Wechselwirkungen in Festkörpern.

Eine sehr geeignete Methode zur Untersuchung von Dynamik in Festkörpern stellt die ^2H NMR dar. Dabei wird die Tatsache ausgenutzt, dass der Kernspin des Deuteriumatoms mit einem elektrischen Quadrupolmoment verbunden ist. Letzteres koppelt an den elektrischen Feldgradienten der chemischen Bindung zum Nachbaratom, wodurch das Deuteriumatom eine exzellente Sonde für die Untersuchung der Reorientierungsdynamik eines Moleküls darstellt. Da die Quadrupol-Wechselwirkung auf den Deuteriumkern beschränkt ist, sind zur Analyse der Spektren und der Relaxationszeitmessungen nur wenige strukturelle Annahmen notwendig. Die Auswirkungen auf Struktur und Dynamik des Systems durch den Austausch von Protonen durch Deuteriumatome sind vernachlässigbar klein. ^2H NMR ermöglicht die Unterscheidung verschiedener Bewegungsarten in einem großen Zeitfenster.

Eine alternative Methode zur Untersuchung der Dynamik ist die Myonen-Spin-Resonanz (μSR) Technik. Das Myon, das in die Probe implantiert wird und als Myoniumatom analog zum Wasserstoffatom beispielsweise an Doppelbindungen addieren kann, dient einerseits als eine Art Sonde, die Informationen über Dynamik, Struktur und chemische Umsetzung liefert. Gleichzeitig wird automatisch ein Radikal generiert, was ein klarer Vorteil dieser Methode ist. Die μSR-Technik stellt eine Variante der magnetischen Resonanz dar, was die theoretischen Grundlagen und die Interpretation der Daten betrifft. Bezüglich der Methode der Erzeugung der Radikale und der Detektion gibt es jedoch erhebliche Unterschiede. Das verwendete Myon gehört zu den Elementarteilchen, trägt den Spin 1/2, besitzt nur 1/9 der Masse sowie das 3.18-fache magnetische Moment des Protons. Es zerfällt mit einer durchschnittlichen Lebensdauer von 2.197 μs unter anderem in ein Positron. Da Myonen an entsprechenden Beschleunigeranlagen mit einer nahezu 100%igen Spinpolarisation erhältlich sind und der Zerfall anisotrop verläuft, ermöglicht die Detektion der Zerfallspositronen Aussagen über die Spinentwicklung in der Probe und damit über die Umgebung des Myons, z.B. das Radikal. In der hier vorliegenden Arbeit kam hauptsächlich eine Methode der μSR, nämlich die Avoided-Level-Crossing (ALC) Technik zum Einsatz. In hohen magnetischen Feldern kommt es in einem 3-spin-1/2-System, beispielsweise aus Proton, Elektron und Myon wie es in einem Mu-substituierten organischen Radikal der Fall ist, zu Überschneidungen der Energieniveaus. Die Kreuzung der Energieniveaus wird jedoch vermieden. Dadurch entstehen Energieaufspaltungen, die sogenannten "avoided level crossings", an denen sich der Spin der beteiligten Teilchen entwickeln kann. Die Darstellung der Zerfallspositronen bezüglich ihres Detektionsortes ergibt dann charakteristische Resonanzen, die Aussagen über die Dynamik erlauben. Dabei wird die Reorientierungsdynamik über die partielle Ausmittelung der Hyperfeinanisotropie eines zum Radikalgerüst in definierter Orientierung stehenden Tensors abgeleitet.

Mittels ALC-μSR wurden in der vorliegenden Arbeit Mu-substituierte Cyclohexadienyl-Radikale, die durch Addition von Myonium an Benzol generiert werden, in metallausgetauschten Zeolithen NaY und NaX untersucht. Benzol bietet sich hier als Adsorbatmolekül an, da das Molekül selber und auch das daraus generierte Radikal sowohl durch μSR als auch durch zahlreiche andere Methoden bereits gut untersucht ist. Ausserdem wird angenommen, dass sowohl das Molekül als auch das Radikal in der Katalyse eine wichtige Rolle spielen. Durch die einfache Struktur ist sichergestellt, dass sich aus dem Benzol nur eine Radikalspezies bildet, was die Analyse der Daten vereinfacht. Als Übergangsmetalle kamen Platin, Palladium, Silber, Nickel und Zink in NaY sowie Mangan in NaX zum Einsatz. Zu Vergleichszwecken wurden auch die nichtausgetauschten Zeolithe NaY und NaX mit Benzol beladen und untersucht.

Ziel war es, die Wechselwirkung der Metallkationen mit den Cyclohexadienyl-Radikalen sowie die Reorientierungsdynamik der Radikale zu bestimmen. Die Auswahl der Metalle wurde dahingehend getroffen, Vergleiche zwischen zweifach und einfach positiv geladenen sowie dia- und paramagnetischen Metallen ziehen zu können. In dieser Arbeit soll außerdem gezeigt werden, wie die Kationendynamik von Pyridiniumtetrafluoroborat und Pyridiniumperchlorat mittels Myonen-Spin-Resonanz und ^{2}H NMR (ausschliesslich am Beispiel des Pyridiniumtetrafluoroborats) aufgeklärt werden kann. Von besonderem Interesse sind dabei die Phasenübergänge, die im Tetrafluoroborat kontinuierlich, im Perchlorat dagegen diskontinuierlich verlaufen. In beiden Methoden, der μSR und der ^{2}H NMR, wird in diesem Zusammenhang ausschliesslich das Pyridiniumkation beobachtet. Daraus ergibt sich bei beiden Methoden die Möglichkeit, durch einen Vergleich mit makroskopischen Daten den Anionenanteil an der entsprechenden physikalischen Messgröße zu bestimmen.

μSR an Zeolithen

Den Zeolithen NaY und NaX liegt die Faujasitstruktur zugrunde. Der einzige Unterschied zwischen den beiden Verbindungen ist das Silizium-zu-Aluminium Verhältnis. Da es für NaX kleiner ist, ergibt sich daraus eine höhere Gerüstladung und damit eine grössere Anzahl an Kationen, die in den Hohlräumen der Gerüststruktur eingebaut sind. Trotzdem sehen die ALC-μSR Spektren der beiden benzolbeladenen Zeolithe sehr unterschiedlich aus.

Die Spektren des benzolbeladenen Zeolithen NaX erinnern stark an Spektren von reinem Benzol. Das lässt auf nahezu ungestörte Cyclohexadienyl-Radikale schliessen. Der einzige Platz in einem Faujasitgerüst, an dem sich die Radikale wie in reinem Benzol verhalten könnten, befindet sich im 12-Ring-Fenster der Superkäfige. Dort gehen die Radikale keinerlei Wechselwirkungen mit Kationen ein. Wieso im NaX keine Wechselwirkung mit den Natriumionen gefunden wird, obwohl deren Zahl deutlich höher ist als in NaY, ist bisher ungeklärt. Die Ergebnisse stimmen jedoch mit Untersuchungen von Böhlmann et al. überein, der chemische Verschiebungen verschiedener Olefine in NaX bei Beladungen mit mehr als einem Molekül pro Superkäfig untersucht hat. Auch hier zeigt sich, dass die Ergebnisse mit denen der reinen Komponenten vergleichbar sind.

NaY beladen mit Benzol wurde mittels μSR bereits von Fleming et al. untersucht. Um einen direkten Vergleich mit den metallausgetauschten Proben zu ermöglichen, wurde C_6H_6/NaY in diesem Zusammenhang auch bei höheren Temperaturen untersucht. Dadurch wurde sichergestellt, dass die Probenpräparation und die Probenform der metallausgetauschten und der nicht-metallausgetauschten Proben identisch sind. Wie zu erwarten zeigen die Spektren von C_6H_6/NaY auf den ersten Blick identische Resonanzen wie die von Fleming *et al.* Drei Resonanzen können als Myonen Δ_1-Linien identifiziert werden. Entsprechend der Zuordnung von Fleming handelt es sich um ein Radikal im 12-Ring-Fenster der Superkäfige und zwei Radikale, die eine Wechselwirkung mit Natriumionen am Platz SII eingehen. Eines davon befindet sich in endo-, das andere in exo-Orientierung zum Metallkation. Während das Radikal im 12-Ring-Fenster eine Myonen-Kopplungskonstante vergleichbar zu der in reinem Benzol zeigt, liegt der Wert für die exo-Orientierung um 20% höher, während der der endo-Orientierung entsprechend niedriger liegt. In den hier vorliegenden Messungen zeigt sich jedoch, dass die Resonanzen für die endo- und die exo-Orientierung Überlagerungen von zwei Linien sein müssen. Eine der beiden Linien weist eine axiale Linienform auf. Dies würde auf eine Rotationsbewegung um die Senkrechte der Molekülebene schließen lassen. Letzteres wurde jedoch durch Rechnungen von Macrae ausgeschlossen, der ein nahezu unbewegliches Cyclohexadienyl-Radikal in NaY gefunden hat. Desweiteren zeigt die andere der beiden Resonanzen eine sehr grosse Intensität und außerdem keine Temperaturabhängigkeit der Hyperfeinkopplungskonstante, wie es die beiden erstgenannten Δ_1-Resonanzen tun. Und schließlich ist auch die Linienform dieser Resonanz nicht eindeutig. Messungen an C_6D_6/NaY haben deutlich die überlagerten Linien gezeigt.

Die den Δ_1-Linien entsprechenden Δ_0-Resonanzen konnten ebenfalls beobachtet und aufgrund ihrer Hyperfeinkopplungskonstante zugeordnet werden. Die Kopplungskonstante der Methylen-Protonen-Resonanz des Radikals im 12-Ring-Fenster ist dabei jedoch grösser als von Fleming vorhergesagt. Eine weitere Δ_0-Resonanz bei relativ hohem Feld konnte über den gesamten Temperaturbereich beobachtet werden. Die Vermutung von Fleming, es könne sich dabei um eine Natrium-Linie handeln, muss jedoch ausgeschlossen werden, da diese Resonanz im Spektrum von C_6D_6/NaY stark verschoben wird. Durch die Deuterierung des Benzols werden die Protonen-Resonanzen im ALC-μSR Spektrum sehr deutlich, die Myonen- und Natrium-Resonanzen dagegen nur geringfügig verschoben. Daraus ergibt sich eine weitere Möglichkeit der Zuordnung der Signale.

Ein Ziel der hier vorliegenden Arbeit war die Identifizierung der Wechselwirkungen von Cyclohexadienyl-Radikalen mit Übergangsmetallkationen in den Zeolithen NaY und NaX. Dazu wurden eine Reihe von metallausgetauschten Zeolithen (C_6H_6/PtNaY, C_6H_6/PdNaY, C_6H_6/NiNaY, C_6H_6/AgNaY, C_6H_6/ZnNaY und C_6H_6/MnNaX) untersucht. Ein direkter Vergleich mit den nichtausgetauschten Zeolithen ist bisher nicht möglich, da die entsprechenden Spektren bisher nur unvollständig interpretiert werden konnten. In der Literatur werden mehrere Beispiele für μSR Spektren von metallausgetauschten Zeoliten beschrieben. Dazu gehören C_6H_6/Li/, C_6H_6/Na/ und C_6H_6/Cu/ZSM-5. Ein typisches Merkmal dieser Spektren ist eine sehr breite und intensive Linie bei tiefem Feld. Diese Resonanz wurde in allen drei Fällen auf eine Metall-Radikal-Wechselwirkung zurückgeführt.

Auch die Spektren von C_6H_6/AgNaY, C_6H_6/ZnNaY und C_6H_6/NiNaY zeigen diese typische Resonanz und sind auch bezüglich der übrigen Resonanzen mit den oben genannten Beispielen vergleichbar was die Linienposition, die Linienbreite und die Intensität betrifft. Berechnet man aus den Linienpositionen die Hyperfeinkopplungskonstanten für die entsprechenden Metalle erhält man sinnvolle Werte. Die daraus berechenten s-Orbital Spindichten bewegen sich in einem Bereich von 2.3% für Nickel bis 3.2% für Silber. Der entsprechende Wert für Kupfer in C_6H_6/Cu/ZSM-5 beträgt 2.8%. Vergleicht man die Werte für Silber mit Ergebnissen von Gee *et al.*, der Silberkomplexe mit Cyclohexadienen (2.8-3.2%) untersucht hat, findet man ebenfalls eine zufriedenstellende Übereinstimmung. Im Gegensatz dazu sind die Spindichten für die Alkalimetalle deutlich grösser (Na: 10%; Li: 14% bzw. 16%). Dies deckt sich jedoch mit Rechnungen von Webster *et al.*, der für eine Wechselwirkung des Natriumions mit einem Cyclohexadienyl-Radikal 12% Spindichte gefunden hat. Eine höhere Spindichte zeigt eine stärkere Wechselwirkung zwischen dem Metall und dem Radikal. Dies ist insofern beachtenswert, als dass die Alkalimetallionen eine schwächere Wechselwirkung zeigen sollten, da sie einen grösseren Ionenradius aufweisen. Dabei bleibt jedoch zu beachten, dass die Metallresonanzen sehr breit sind. Da die Spindichten für Natrium und Lithium aus den Rohspektren bestimmt wurden, ist der Fehler dementsprechend gross.
Alle metallausgetauschten Proben, die eine Metallresonanz zeigen, zeigen keine Natriumresonanzen, obwohl erreichbare Natriumionen in den Zeolithen vorhanden sind. Daraus kann geschlossen werden, dass die Cyclohexadienyl-Radikale die Übergangsmetallkationen bevorzugen. Für keine der gefundenen Metall-Radikal-Wechselwirkungen wurden Resonanzen entdeckt, die auf verschiedene Orientierungen des Radikals wie von Fleming *et al.* für C_6H_6/NaY vorgeschlagen schließen lassen. Gleichzeitig weisen alle metallausgetauschten Proben, die eine Metallresonanz zeigen, eine Δ_1-Resonanz von nur sehr geringer Intensität auf. Die deutet auf eine zusätzliche Bewegung hin. Es wäre möglich, dass die Cyclohexadienyl-Radikale innerhalb eines Superkäfigs zwischen den vier tetraedrisch angeordneten Natriumionen springen.

C_6H_6/PtNaY zeigt keine Metallresonanz. Die übrigen Resonanzen lassen jedoch auf ein nahezu ungestörtes Cyclohexadienyl-Radikal schliessen. Da die Elementaranalyse der Probe eine erhebliche Menge an Stickstoff gezeigt hat, muss davon ausgegangen werden, dass die Kalzinierung der Probe nicht vollständig stattgefunden hat. Die Platinionen werden vermutlich von den Aminliganden abgeschirmt. So wird eine Wechselwirkung mit dem Radikal unmöglich gemacht. Die Benzolmoleküle befinden sich stattdessen in den 12-Ring-Fenstern der

Superkäfige. Da alle Resonanzen sehr klein sind, muss außerdem davon ausgegangen werden, dass zumindest ein Teil der Benzolmoleküle durch die Platinionen oxidiert worden ist. Für C_6H_6/PdNaY konnten überhaupt keine Resonanzen detektiert werden. Die Analyse der Probe hat gezeigt, dass während der Probenpräparation Redoxprozesse stattgefunden haben müssen, die das Benzol komplett zerstört haben. Möglicherweise fungiert das Palladium in dieser Probe als wirksamer Katalysator oder ist selber am Redoxprozess beteiligt. C_6H_6/MnNaX hat ebenfalls keine Resonanzen gezeigt. Mangan ist das einzige paramagnetische Metallion, das in der vorliegenden Arbeit verwendet wurde. Es ist davon auszugehen, dass durch eine Wechselwirkung des Radikalelektrons mit den ungepaarten Elektronen des Metallions Relaxationsprozesse stattfinden, die das Auftreten von μSR-Resonanzen verhindern.

μSR an Pyridinium-Salzen

Pyridiniumtetrafluoroborat und Pyridiniumperchlorat bestehen aus identischen Kationen, unterscheiden sich jedoch bezüglich des Anions. Die beiden Verbindungen sind vergleichbar was die Kristallstruktur und die Änderung der Kristallsymmetrie mit der Temperatur betrifft. Sie zeigen jedoch deutliche Unterschiede in der Ordnung der auftretenden Phasenübergänge. Das Tetrafluoroborat zeigt zwei Phasenübergänge zweiter Ordnung, das Perchlorat dagegen zwei Übergänge erster Ordnung. Beide Verbindungen werden jedoch unterhalb des Hochtemperaturüberganges ferroelektrisch. Es wird erwartet, dass beide Verbindungen vergleichbare μSR-Spektren bezüglich der Resonanzposition und der Linienform liefern. Dagegen sollte die Temperaturabhängigkeit der Spektren aufgrund der verschiedenartigen Ordnungen der Übergänge deutliche Unterschiede aufweisen. Myonium sollte analog zur Addition an einen Benzolring an das Pyridiniumkation addieren können, das sich lediglich durch ein Stickstoffatom anstelle eines Kohlenstoffatoms im Ring vom Benzol unterscheidet. Daher resultiert auch die positive Ladung des Rings. Bei der Addition von Myonium an das Pyridiniumkation können prinzipiell vier verschiedene Radikale entstehen, eines durch die Addition direkt am Stickstoff (ipso-Radikal) und drei durch die Addition in ortho-, meta- und para-Stellung zum Stickstoff (ortho-, meta- und para-Radikal). Jedes dieser Radikale liefert prinzipiell zwei intensive Resonanzen (Myon und Methylen-Proton) und einige weniger intensive Linien, die auf die jeweiligen Ringprotonen zurückzuführen sind.

Das ALC-μSR Spektrum von Pyridiniumtetrafluoroborat bei Raumtemperatur zeigt sechs intensive Resonanzen, drei weniger intensive Signale und etliche kaum erkennbare Linien. Transversalfeldmessungen konnten die Vermutung bestätigen, dass die sechs intensiven Signale von drei verschiedenen Radikalen stammen. Ein Vergleich der daraus berechneten Kopplungskonstanten mit den Werten für aza-Cyclohexadienyl-Radikale in Poylvinylpyridin von Ronfard Haret *et al.* erlaubt die Zuordnung der einzelnen Signale zum ortho-, meta- und para-Radikal. Das ALC-μSR Spektrum des Pyridiniumperchlorats bei Raumtemperatur ist wie erwartet identisch mit dem des Tetrafluoroborats. Das Auftreten von Δ_1-Resonanzen bei beiden Verbindungen zeigt, dass über den gesamten Temperaturbereich anisotrope Bedingungen vorliegen. Alle intensiven Signale werden mit zunehmender Temperatur schmaler. Dies deutet auf eine zunehmende dynamische Ausmittelung hin.

Im Gegensatz dazu ist die Entwicklung der Linienformen mit der Temperatur in den beiden Verbindungen deutlich verschieden. Zur Beurteilung der Temperaturabhängigkeit der Linienform werden die D_{zz}-Werte herangezogen. Sie sind vergleichbar mit der Linienbreite von Lorentz-Linien und erlauben Aussagen über die Anisotropie des Systems. Die Betrachtung wird für beide Verbindungen auf das ortho-Radikal beschränkt und am Beispiel der zugehörigen Myonen-Resonanz (Δ_1) dargelegt, da Veränderungen in der Anisotropie den grössten Einfluss auf Δ_1-Resonanzen zeigen. In der Hochtemperaturphase sind die D_{zz}-Werte der Δ_1-Resonanz des ortho-Radikals für beide Salze identisch. Daraus folgt, dass die Dynamik der Kationen in dieser Phase unabhängig vom Anion ist. An dem negativen Vorzeichen der D_{zz}-Werte lässt sich erkennen, dass das aza-Cyclohexadienyl-Radikal in beiden Verbindungen eine schnelle Rotationsbewegung um die Achse senkrecht zur Moleküleben ausführt. Am Hochtemperaturphasenübergang des Tetrafluoroborates verändert sich die Linienbreite quasi in einem S-förmigen Verlauf, der an den in der Literatur propagierten kontinuierlichen Phasenübergang erinnert. Die D_{zz}-Werte der entsprechenden Resonanz im Perchlorat ändern sich bei der entsprechenden Temperatur dagegen stufenförmig. Dies entspricht einem diskontinuierlichen Phasenübergang. Im Perchlorat wurde im Gegensatz zum Tetrafluoroborat außerdem eine Unterkühlung festgestellt, die ebenfalls auf einen Übergang erster Ordnung schließen lässt. Die Linienverbreiterung der Δ_1-Resonanzen unterhalb der Hochtemperaturübergänge ist in beiden Fällen so drastisch, dass eine Untersuchung der Tieftemperaturübergänge mittels ALC-μSR nicht möglich war.

NMR an Pyridinium-Salzen

Die Dynamik des perdeuterierten Pyridiniumkations in Pyridiniumtetrafluoroborat wurde mittels dynamischer ^{2}H NMR untersucht. Dabei kamen sowohl Linienformanalysen (Quadrupol-Echo-Experimente) als auch Spin-Gitter-Relaxationszeitmessungen (Inversion-Recovery-Experimente) zum Einsatz. Besondere Bedeutung kam den beiden fest-fest Phasenübergängen bei 204 K und 238,7 K zu, die in der Literatur als kontinuierliche Übergänge beschrieben werden. Die Methode berücksichtigt lediglich die Kationen des Systems, da das Anion keine Deuteriumatome enthält und somit unsichtbar bleibt.

Die experimentellen Spektren zeigen für das Kation in der Hochtemperaturphase eine schnelle Rotationsbewegung um die pseudo $C6$-Achse. Mit sinkender Temperatur wird diese Rotation mehr und mehr behindert, bis das Kation bei 120 K schliesslich fast unbeweglich und in einer Vorzugsorientierung ausgerichtet verbleibt. Ein einfaches 2-Sprung-Modell, das einen Sprungprozess mit variierendem Sprungwinkel zwischen zwei gleichbesetzten Orientierungen beschreibt, ist in der Lage die Quadrupol-Echo-Experimente zufriedenstellend zu simulieren. Die Spin-Gitter-Relaxationszeitmessungen können damit jedoch nicht erklärt werden. In Anlehnung an die $R\bar{3}m$ Symmetrie des Pyridiniumtetrafluoroborats in der Hochtemperaturphase wurden sowohl ein 3-Sprung-Modell als auch ein 6-Sprung-Modell gewählt. Das 3-Sprung-Modell geht davon aus, dass sich das Pyridiniumkation in drei verschiedenen Orientierungen befinden kann, die jeweils 120° auseinander liegen. Dementsprechend können im 6-Sprung-Modell sechs verschiedene Orientierungen (60° Sprünge) besetzt werden. In beiden Modellen werden den Orientierungen Besetzungswahrscheinlichkeiten zugeordnet. Eine Veränderung der Besetzung führt zu einer Veränderung der Linienform. Die Simulation der Relaxationszeitmessungen erfolgt mit diesen Besetzungen aus der Linienformanalyse und der Variation der Korrelationszeit der Sprünge. Als Ergebnis wird die T_{1Z}-Relaxationszeit erhalten. Die optimale Simulation wurde sowohl bei der Linienformanalyse als auch bei den Relaxationszeitmessungen durch einen optischen Vergleich der überlagerten experimentellen und simulierten Spektren erreicht. Sowohl die Quadrupol-Echo- als auch die Inversion-Recovery-Experimente können mit beiden Simulationsmodellen gleichermaßen zufriedenstellend simuliert werden. Die Übereinstimmung der experimentellen und der simulierten Spektren für beide Experimente zeigt, dass sowohl das 3-Sprung-Modell als auch das 6-Sprung-Modell geeignet ist, die Dynamik des Pyridiniumkations in Pyridiniumtetrafluoroborat zu beschreiben. Es sollte jedoch beachtet werden, dass die Übereinstimmung zwischen Simulation und Experiment nicht bedeutet, dass es sich bei der untersuchten Dynamik unweigerlich um einen Mechanimus entsprechend dem Modell handeln muss; das Modell kann lediglich nicht ausgeschlossen werden. Stimmen Simulation und experimentelle Daten auf der Basis des gewählten Modells jedoch nicht überein, kann letzteres eindeutig ausgeschlossen werden.

Die Besetzungswahrscheinlichkeiten aus den Simulationsmodellen sind in der Tieftemperaturphase für beide Modelle identisch, in der mittleren Temperaturphase nahezu identisch. In der Hochtemperaturphase findet man für beide Modelle Gleichbesetzung aller Orientierungen. Der Verlauf der Besetzungswahrscheinlichkeiten mit der Temperatur zeigt, dass mit abnehmender Temperatur eine Vorzugsausrichtung gegenüber den anderen Orientierungen deutlich an Bedeutung gewinnt und bei 120 K fast den Wert eins erreicht. Dementsprechend sind bei tiefer Temperatur fast alle Kationen in dieser Orientierung ausgerichtet.

Das Minimum der experimentellen T_{1Z}-Kurve in der mittleren Temperaturphase deutet darauf hin, dass die Korrelationszeit der beobachteten Bewegung in der Größenordnung der Larmor-Frequenz von $2\pi \times 46{,}07$ MHz liegt. Ein Vergleich mit den simulierten T_{1Z}-Werten zeigt, dass der Verlauf qualitativ identisch ist. Die Korrelationszeiten verringern sich mit abnehmender Temperatur. In allen drei Phasen erkennt man einen linearen Verlauf. Daraus ergeben sich für das 3-Sprung-Modell Aktivierungsenergien von 10,0 kJ mol^{-1}, 46,4 kJ mol^{-1} und 19,3 kJ mol^{-1} für die Tief-, die mittlere und die Hochtemperaturphase. Für das 6-Sprung-Modell ergeben sich ähnliche Werte. Die Aktivierungsenergien sind insofern erstaunlich, als sie ein Maximum in der mittleren Temperaturphase zeigen.

Aus den erhaltenen Besetzungswahrscheinlichkeiten der drei Orientierungen können die Polarisation und die Enthalpieänderung bezüglich des Kations berechnet werden. Ein Vergleich der Ergebnisse aus der Simulation mit den Resultaten aus makroskopischen Messungen zeigt, dass die Werte deutliche Unterschiede aufweisen. Lediglich in der mittleren Temperaturphase findet man nahezu identische Werte für das 3-Sprung-Modell. Daraus lässt sich folgern, dass die Orientierung des Kations nicht der einzige Mechanismus ist, der zur Polarisation und zur Enthalpie beiträgt. Es muss ein zusätzlicher Mechanismus beitragen, der einen kompensatorischen Effekt ausübt. Dies könnte der klassische ferroelektrische Mechanismus sein, bei dem die Ionenuntergitter gegeneinander verschoben werden. Eine Verschiebung von Anionen- und Kationengitter von 0,23-0,25 Å würde ausreichen, um die fehlende Polarisation auszugleichen. Ein Abschätzung der Oszillationsfrequenz in der Tieftemperaturphase zeigt, dass die Werte im Bereich der experimentellen Ergebnisse liegen.

Beide Mechanismen, sowohl die Orientierung der Kationen als auch der klassische ferroelektrische Mechanismus, sind kontinuierlich, was den Angaben in der Literatur bezüglich der Ordnung der Phasenübergänge in Pyridiniumtetrafluoroborat entspricht; allerdings ist es wahrscheinlicher, dass es sich bei beiden Mechanismen um Grenzfälle zwischen erster und zweiter Ordnung handelt. Die Linienformanalyse der ALC-μSR-Spektren zeigt für das Tetrafluoroborat einen Hochtemperaturübergang zweiter Ordnung. Ebenso weist die Linienform in der DSC für beide Übergänge auf zweite Ordnung hin. Außerdem wurden für das Perchlorat sowohl in den DSC Messungen als auch in den ALC Spektren eindeutig Hinweise auf Unterkühlung gefunden, die beim Tetrafluoroborat fehlen. Dies weist ebenfalls auf kontinuierliche Phasenübergänge in Pyridiniumtetrafluoroborat hin. Im Gegensatz dazu zeigt die ferroelektrische Polarisationskurve von Pyridiniumtetrafluoroborat jedoch ein kritisches Verhalten, das deutlich näher an einem Phasenübergang erster denn zweiter Ordnung liegt. Die Resultate der hier vorliegenden Arbeit stehen in deutlichem Kontrast zu den Ergebnissen von Hanaya *et al.*, der den Ursprung der Ferroelektrizität in Pyridiniumtetrafluoroborat auf die Ordnung der Tetrafluoroboratanionen zurückführt. Es ist zwar durchaus sinnvoll, dass die Unordnung der Anionen signifikant zur Wärmekapazität und zur Entropie beiträgt; zur Ferroelektrizität können die Anionen jedoch nur beitragen, wenn deren Tetraedersymmetrie deutlich verzerrt ist.

Eine aktuelle Strukturanalyse des Tetrafluoroborates in der Tieftemperaturphase könnte hier weiterhelfen, die Daten zu interpretieren. Desweiteren wären vergleichbare ^{2}H NMR-Untersuchungen an Pyridiniumperchlorat sicherlich aufschlussreich.

Ein direkter Vergleich der Ergebnisse aus den μSR-Messungen an Pyridinium-tetrafluoroborat mit denen aus den ^{2}H NMR-Messungen ist schwierig, da die beiden Techniken offensichtlich verschiedene Zeitskalen abdecken. Daraus resultiert jedoch, dass die beiden Methoden komplementäre Ergebnisse liefern können. Dabei muss beachtet werden, dass in den μSR-Messungen das aza-Cyclohexadienyl-Radikal beobachtet wird, in den NMR-Messungen jedoch das perdeuterierte Pyridiniumkation. μSR-Messungen an Pyridiniumperchlorat haben jedoch gezeigt, dass die Methode gut geeignet ist, kollektive Eigenschaften zu untersuchen. So wurde nicht nur der Phasenübergang erster Ordnung sondern auch eine Unterkühlung festgestellt. Dieses Phänomen ist typisch für diskontinuierliche Übergänge und wurde in PyBF$_4$ dementsprechend nicht beobachtet. Mit beiden Methoden wird in der Hochtemperaturphase des PyBF$_4$ eine schnelle Rotationsbewegung des Kations um die Achse senkrecht zur Molekülebene beobachtet. Die μSR Spektren zeigen deutlich eine zusätzliche Ausmittelung der Anisotropie durch eine Bewegung der Rotationsachse selber, die in den NMR-Messungen nicht festgestellt werden kann. Mit abnehmender Temperatur kann in beiden Methoden übereinstimmend beobachtet werden, dass die Rotationsbewegung mehr und mehr eingeschränkt wird, wobei die deutlichsten Veränderungen an den Phasenübergängen auftreten. In der μSR tritt in der mittleren Phase eine derartige Linienverbreiterung ein, dass eine Untersuchung des Tieftemperaturüberganges unmöglich wird. Hier liegt ein deutlicher Vorteil der ^{2}H NMR, mit der PyBF$_4$ in allen drei Phasen sehr gut untersucht werden kann. Die ALC-μSR Messungen haben gezeigt, dass es möglich ist, mit dieser Methode die Ordnung der Phasenübergänge zu unterscheiden. So zeigt die Veränderung der Linienform am Hochtemperaturphasenübergang für das PyBF$_4$ einen S-förmigen, für das PyClO$_4$ dagegen einen stufenförmigen Verlauf entsprechend einem Übergang zweiter bzw. erster Ordnung. Das Perchlorat wurde mittels ^{2}H NMR bisher nicht untersucht. Dies wäre insofern interessant, als dass bereits das Tetrafluoroborat einen eher stufenförmigen Verlauf der Besetzungswahrscheinlichkeiten der verschiedenen Orientierungen aufweist, obwohl der Phasenübergang kontinuierlich verläuft. Die Simulation der NMR-Spektren ermöglicht die Ableitung thermodynamischer und kinetischer Daten. Da sich diese Daten nur auf die Dynamik des Pyridiniumkations beziehen, ergibt sich daraus die Möglichkeit durch einen Vergleich mit makroskopischen Daten die einzelnen Ionenanteile zu bestimmen. Dabei muss jedoch ausdrücklich darauf hingewiesen werden, dass die gesamte Auswertung der NMR-Daten auf dem der Simulation zugrunde liegenden mechanistischen Modell beruht. Da eine Übereinstimmung der simulierten mit den experimentellen Spektren das entsprechende Modell lediglich nicht ausschließt, aber nicht eindeutig beweist, kann daraus nicht die absolute Richtigkeit der Interpretation abgeleitet werden.

DANK

Ohne die Unterstützung vieler Menschen in meinem Umfeld wäre diese Arbeit nie zustande gekommen.

Mein ganz besonderer Dank gilt Herrn Prof. Dr. Emil Roduner für die Aufnahme in seine Arbeitsgruppe sowie die wissenschaftliche Betreuung meiner Arbeit, die sich durch sein ununterbrochenes Interesse am Fortgang der Arbeit, durch viele Anregungen und durch eine unendliche Diskussionsbereitschaft auszeichnete.
In großer Schuld stehe ich bei Herrn Prof. Dr. Klaus Müller, der die Betreuung des NMR-Teils dieser Arbeit übernommen hat, indem er nicht nur das Spektrometer und die zugehörigen Auswertungsprogramme zur Verfügung gestellt, sondern auch sein Fachwissen und seine Erfahrung auf diesem Gebiet in die Arbeit eingebracht hat.
Herrn Prof. Dr. Thomas Schleid bin ich für die Übernahme des Prüfungsvorsitzes und Herrn PD Dr. Günter Majer für die freundliche Übernahme des Zweitberichtes zu Dank verpflichtet.

Herrn Prof. Dr. Don G. Fleming aus Vancouver, Kanada, danke ich sehr herzlich für seine theoretische und praktische Unterstützung bei den Messungen am PSI und der anschließenden Auswertung der Daten. Seine Arbeiten an NaY waren die Grundlage für die hier präsentierten Ergebnisse.
Herr Dr. Piotr Czarnecki aus Poznan, Polen, hat die Untersuchung der Pyridinium-Salze mittels μSR angeregt. Ohne seine Initiative wäre ein großer Teil dieser Arbeit nie zustande gekommen. Ihm danke ich für die Unterstützung bei der Interpretation der Ergebnisse.

Ohne die Hilfe der Mitarbeiter am PSI wären uns die Messungen unmöglich gewesen. Hier möchte ich besonders Herrn Dr. Dierk Herlach erwähnen. Herrn Dr. I. D. Reid, Herrn X. Donath und Herrn Dr. N. Suleimanov danke ich für die technische Unterstützung.

Die Kollegen meiner Arbeitsgruppe haben für ein sehr gutes Arbeitsklima gesorgt und zu dieser Arbeit beigetragen. Besonders zu erwähnen sind hier Herr Dr. Herbert Dilger und Herr Dr. Robert Scheuermann, ohne deren Hilfe die Messungen am Paul-Scherrer-Institut nicht möglich und nur halb so erträglich gewesen wären. Ohne sie wäre ich auch an etlichen Computerproblemen gescheitert.
Mein Dank gilt Herrn Dr. Thorsten Schmauke für die Einweihung in die Geheimnisse der Zeolith-Synthese, Herrn Dr. Georg Hübner für die Hilfe bei den Rechnungen sowie Herrn Dipl.-Chem. Holger Bühner für die Einführung in das Institut.

Ohne die Betreuung von Frau Inge Blankenship wäre ich vermutlich an der Administration kläglich gescheitert. Ihr gilt mein ganz persönlicher Dank für so manches nette Wort, mit dem sie maßgeblich zum guten Betriebsklima beigetragen hat.

Aber auch Betreuern und Kollegen aus den anderen Arbeitsgruppen des Instituts für Physikalische Chemie und aus dem Graduiertenkolleg "Magnetische Resonanz" sei an dieser Stelle herzlich gedankt. Insbesondere Herrn Dipl.-Chem. Gerald Fritsch und Herrn M. Sc. Jorge Antonio Villanueva-Garibay aus der Arbeitsgruppe von Herrn Prof. Dr. K. Müller danke ich für die Einführung in den Umgang mit dem NMR-Spektrometer.

Der mechanischen, der elektronischen und der Glasbläser-Werkstatt sowie der Chemotechnik sei an dieser Stelle sehr herzlich für die Unterstützung gedankt. Besonders möchte ich an dieser Stelle Herrn Peter Haller aus der mechanischen Werkstatt erwähnen, der sich auch durch die fast unendliche Undichtigkeit der Zink-Zelle nie hat entmutigen lassen.

Für die Analyse der Zeolith-Proben möchte ich Frau Förtsch aus der Anorganischen Chemie (IAC) sowie Frau D. Göhringer und Frau C. Lauxmann aus der Organischen Chemie (IOC) (Elementaranalyse) sowie Herrn B. Janisch vom Institut für Verfahrenstechnik und Dampfkesselwesen (IVD)(RFA) sehr herzlich danken.

Der Deutschen Forschungsgemeinschaft (DFG) bin ich für die finanzielle Unterstützung im Rahmen des Graduiertenkollegs *Moderne Methoden der Magnetischen Resonanz in der Materialforschung* an der Universität Stuttgart zu Dank verpflichtet.

Meiner Mutter und meinem Vater gilt mein ganz besonderer Dank dafür, dass sie mir den Mut gegeben haben, mich auf diese Reise zu begeben. Sie haben mich auf dem gesamten Weg begleitet und nie den Glauben an mich und mein Ziel verloren. Jörg danke ich ganz herzlich dafür, dass er in dieser Zeit mit mir alle Höhen und Tiefen durchlebt hat.

Hartmut danke ich für die mentale Unterstützung besonders in der Endphase dieser Dissertation und für das Korrekturlesen. Stefan danke ich für die LaTeX-online-Hilfe.

Und Sandra schliesslich hat dafür gesorgt, dass ich mit meinen zwei Füßen immer auf dem Boden der Tatsachen geblieben bin. Ihr habe ich diese Arbeit gewidmet.

DANKE

ACKNOWLEDGEMENT

Without the active support of many people this thesis would not have been accomplished.

First of all I want to thank Prof. Dr. Emil Roduner for giving me the opportunity to collaborate in his group. I am grateful for his interest and active support, all his ideas, invaluable discussions and his dedication.
I am indebted to Prof. Dr. Klaus Müller. He did not only provide the NMR spectrometer and the software for analysing the spectra but he coached also the NMR part of this work and contributed considerably with his knowledge and his experience in this field.

Taking the chair of the examination board by Prof. Dr. Thomas Schleid is kindly acknowledged. Writing the second advisory opinion by PD Dr. Günter Majer is also kindly acknowledged.

Thank you to Prof. Dr. Don G. Fleming from Vancouver, Canada, for his active support at PSI and the fruitful discussions concerning the μSR measurements on the zeolites. The present work is based on his initial investigations on C_6H_6/NaY. And so to Dr. Piotr Czarnecki from Poznan, Poland, for initiating the μSR measurements on the pyridinium salts and for his support to the interpretation of the results, especially for providing the original pyroeffect data.

I would have been lost upon the measurements at PSI without the help of the people from the μSR facility. I would like to mention escpecially Dr. Dierk Herlach. Thank you also to X. Donath, Dr. I. D. Reid and Dr. N. Suleimanov for technical support.

All my colleagues contributed to a pleasant working atmosphere. I would like to mention especially Dr. Herbert Dilger and Dr. Robert Scheuermann and thank them for their active support at PSI. Without their assistance the beam times at PSI would have been only half as pleasant. Besides, they solved a lot of computer problems for me.
Thank you to Dr. Thorsten Schmauke for the assistance in zeolite preparation, to Dr. Georg Hübner for the assistance with the calculations and to Dipl.-Chem. Holger Bühner for introducing me to the institute.
A very special thank you to Mrs. Inge Blankenship who accounted for all the administration. Besides, she contributed to a pleasant working atmosphere significantly.

Thank you to all the people from other groups and institutes and from the Graduate College *Magnetic Resonance*, especially Dipl.-Chem. Gerald Fritsch and M. Sc. Jorge Antonio Villanueva-Garibay for introducing me to the handling of the NMR spectrometer.

The support by the mechanical, the electronical and the glassblower workshop as well as the chemical technicians is kindly acknowledged. Special thanks to Peter Haller from the mechanical workshop who built the μSR sample cells for me and who never gave up although the Zn-cell was really a special case.
The EA by Mrs. Förtsch from the Institute of Inorganic Chemistry (IAC) and the EA by Mrs. D. Göhringer and Mrs. C. Lauxmann from the Institute of Organic Chemistry (IOC) are kindly acknowledged; just as the RFA by Mr. B. Janisch from the IVD.

Financial support by the DFG (Deutsche Forschungsgemeinschaft; Graduate College *Modern Methods of Magnetic Resonance in Materials Science*) is gratefully acknowledged.

I am indebted to my mum and my dad. They encouraged me to go my own way. They accompanied me all the time and they always believed in me and my aim. Thank you to Jörg for bearing all the ups and downs.
Special thanks to Hartmut for mental benefit especially during the endgame of this work and for proofreading. And thank you to Stefan for providing LaTeX online help.
And last but not least, Sandra. She saved me from living in an ivory tower. I dedicated this work to her.

THANK YOU